Pagnoni   Project Engineering

Anastasia Pagnoni

Project Engineering

Computer-Oriented Planning
and Operational Decision Making

With 102 Figures

Springer-Verlag
Berlin Heidelberg New York London Paris
Tokyo Hong Kong Barcelona

Anastasia Pagnoni
University of Modena
Department of Pure and Applied Mathematics
Via Campi, 213/b, I - 41100 Modena, Italy

Present address:
University of Milan
Department of Computer Science
Via Moretto da Brescia, 9, I-20133 Milano, Italy

ISBN-13: 978-3-642-75632-0   e-ISBN-13: 978-3-642-75630-6
DOI: 10.1007/978-3-642-75630-6

Library of Congress Cataloging-in-Publication Data
Pagnoni, Anastasia.
   Project engineering: computer-oriented planning and operational decision making/
Anastasia Pagnoni.
   p. cm.
   Includes bibliographical references and index.

   1. Engineering – Management. 2. Decision-making.   I. Title.
TA194.P34   1990   90-10039   658.4'04 – dc20

© Springer-Verlag Berlin Heidelberg 1990
Softcover reprint of the 1st edition 1990

2145/3140(3011)-543210 – Printed on acid-free paper

To Elena and Francesco,

my marvellous kids

Preface

Project engineering is the discipline concerned with the application of science to the life-cycle of those complex phenomena we call projects.

Very often coordinated human activity is conceived as a project and executed to achieve projected goals. This holds true in every field of human activity: there are manufacturing and building projects, engineering and research projects, organizational and political projects.

In everyday language - and in this book - the word *project* is used to indicate a past, ongoing, or future planned effort, driven by a definite interest, and usually involving a number of people and activities: planned effort for building a house, running a manufacturing process or a scientific experiment, performing a specific unit of research, or organizing a school system.

Project engineering encompasses a multitude of methodologies, and its practice requires familiarity with different techniques and a critical understanding of diverse approaches.

This book offers a unitary presentation of several quite different planning methodologies. Unavoidably, the selection — graphs, activity networks, CPM, PERT, GERT, condition/event nets, place/transition nets, predicate/transition nets — reflects the author's view of project planning. But I am convinced that this choice offers a solid platform for developing and analyzing project plans. In addition, the book introduces operational decision making, a methodology — based on the theory of fuzzy sets — for supporting plan execution supervisors in making operational decisions. The methodologies presented arose partly in the field of computer science and partly in the field of operations research. All of them are here introduced and discussed with the idea that computer tools can be acquired or developed to support planning, analysis and supervision of projects.

I have written this book as a basic introduction to project engineering. I have sought as well to supply an efficient project engineering toolbox, and to offer a unifying introduction to a wide cross-section of the relevant questions and results in the area. Particular care has been devoted to providing practical guidelines on when each of the techniques may be appropriately used. The appendix "Further Reading" will help the reader pursue particular topics in greater depth.

The book is dedicated to professionals and practitioners, but it may well be used as complementary textbook for a graduate level course in system engineering, control engineering, computer aided planning, decision support systems, or operations research.

The subject matter is expounded in a self-contained fashion, though the reader is supposed to possess the mathematical background of a graduate level computer science student.

Acknowledgements

Some features of this book which the author particularly likes were helpful suggestions of Dr. Anatol Holt. Without his thorough reading of the manuscript, the book would have been poorer in content and less pleasant in form. Our lively discussions made writing this book a challenging experience.

A particular acknowledgement to Prof. Luigi Pojaga for revising the exposition of networking techniques with competence and friendly care. Thanks also to Mr. J. Andrew Ross of Springer-Verlag for correcting the English and suggesting substantial improvements of the manuscript.

A special "grazie!" to Dr. Hans Wössner and Mrs. Ingeborg Mayer for their encouraging support and effective advice.

Finally, the author would like to mention her debt to Prof. Giovanni Degli Antoni for having started her in the field of Computer Science, and sustained her ever since.

Milan, January 1990   Anastasia C. Pagnoni

Contents

Mathematical Notations

Sets are denoted by upper case letters, set elements by lower case letters. Vector and matrix identifiers are bold-face, and so are identifiers of relations, graphs, and nets. The following notations are frequently used in the book:

$=$	equal to
$\neq$	different from
$\in$	belongs to
$\notin$	does not belong to
$\supseteq$	comprehends
$\supset$	strictly comprehends
$\{.....\}$	set
$\emptyset$	empty set
$\mathbb{P}(X)$	the power set of X
$\cap$	intersection of subsets
$\cup$	union of subsets
$\setminus$	set difference
$\times$	set product
$\vee$	or
$\wedge$	and
$\neg$	not
$\parallel X \parallel$	the cardinality of set X
$\mathbb{M}(X)$	the set of multisets over set X

$\oplus$	the separator within formal polynomials
$\langle x_1, x_2, \ldots, x_n \rangle$	an n-tuple, or tuple, over a given set
X^n	the set of n-tuples over set X
$\langle \rangle$	the zero-tuple, also called a token
$\exists$	there exists a . . .
$\forall$	for all . . .
$\mid$	such that
$\rightarrow$	if . . . then . . .
$\leftrightarrow$	. . . if and only if . . .
N	the set of naturals
$\Im$	the set of integer numbers
$\Im^+$	the set of non-negative integers
$\Re$	the set of real numbers
$\Re^+$	the set of non-negative real numbers
$\aleph_0$	the cardinality of the set of naturals
Σ	sum
Π	product
$P\ [X]$	the probability of event X
$\mathbf{0}$	matrix of zeros of a given order
$\mathbf{A}^{\mathsf{T}}$	the transpose of matrix $\mathbf{A}$

For the reader's convenience we recall the definition of *multiset,* and some related notions.

Given a finite set $X = \{x_1, x_2, \ldots, x_n\}$, we call every mapping of X into $\Im^+$ a *finite multiset* — or, simply, a *multiset* — over X.

A multiset M over X may be represented by a formal polynomial

$$M(x_1)\ x_1 \oplus M(x_2)\ x_2 \oplus \ldots \oplus M(x_n)\ x_n$$

where $M(x_i)$ is the image of x_i, and sign $\oplus$ is to be regarded as a separator. By convention, coefficients equal to one, and items with coefficients equal to zero are omitted.

M is often interpreted as the collection of $M(x_1)$ copies of item x_1, $M(x_2)$ copies of item x_2, \ldots, and $M(x_n)$ copies of item x_n.

The multiset which associates the number zero with each element of X, is called the *empty multiset*   and denoted by Ø.

Given two multisets over the same finite set $X = \{x_1, x_2, \ldots, x_n\}$

$$M' = M'(x_1)\, x_1 \oplus M'(x_2)\, x_2 \oplus \ldots + M'(x_n)\, x_n \quad \text{and}$$
$$M'' = M''(x_1)\, x_1 \oplus M''(x_2)\, x_2 \oplus \ldots \oplus M''(x_n)\, x_n$$

their *union* M'∪M'' is defined to be the multiset over X

$$M(x_1)\, x_1 \oplus M(x_2)\, x_2 \oplus \ldots \oplus M(x_n)\, x_n \quad \text{with} \quad M(x_i) = M'(x_i) + M''(x_i).$$

We write

$$M' \leq M'' \quad \text{if and only if} \quad \forall\, x \in X: \; M'(x) \leq M''(x).$$

Introduction

Conceiving projects and devising plans are characteristic expressions of any goal-directed human activity. Manufacturing processes, scientific experiments, research programs, tax collection systems, are all endeavors once born as projects — of quite different nature, indeed, but all sharing essential features.

Project engineering is the application of science to the development and analysis of project plans, and to the supervision of their realization.

In project engineering, the word *project* has a very general meaning: it designates any piece of planned work joined by a definite driving interest — past, ongoing or future work, generally involving a number of people and activities, such as the work planned to produce a car, to organize a medical treatment protocol, or to look for new sources of energy.

Project specifications describe the work to be planned in order to put the project's impelling interest into effect. We shall call project specification every consistent description of the work to be done in order to put the impelling interest into effect. Such a description may be provided in a large variety of languages, and in several ways. Initial draft outlines of goals, resources and activities, preparatory sketches, and detailed work plans — all of them are project specifications, indeed all may be different specifications of the same project.

It is a common practice to formulate project specifications in natural language, as a collection of statements about interests and goals, time and resource constraints to be respected, and risk levels to be taken into account. This collection of statements must be checked for consistency and feasibility: it must be *validated*.

A project specification is the formulation of a problem. *Project design* consists in thinking out, and then representing, one feasible solution of that problem. Such a solution is called a *plan*.

There are *planning languages* — languages for representing project plans — of quite different kinds. These are usually more technical than the specification language, for the reason that the specification language is closer to commitment than the planning language.

Plans are global or partial outlines of a specific strategy for carrying out the work intended by a project. A global plan will be a collection of many partial plans, each one relating to some aspects of the devised realization: plans for scheduling time and resources, cost plans, financial plans, plans for controlling executions, etc.

Project verification is the checking of the correspondence between plans and specifications. A positive outcome of this check is a necessary condition for starting any execution of the project. A negative outcome forces partial or global redesigning of the plan. The most broadly applied verification techniques are based on the statistical analysis of plan simulations, and the computation or estimation of time, cost, utility, risk, and parameter values.

The life of a project evolves from an initial, informal set of ideas to the final executable plan, passing through many subsequent intermediate stages, characterized by proper specifications and involving an increasing amount of actual information. At every stage a complex of plans — each referring to one or more realization aspects — are generated from the plans of the prior stage on the basis of supplementary specifications.

The adopted plan representation language will condition the planner with its features and constraints. Also, different languages in general allow for different analysis possibilities. Therefore, choosing a suitable representation language is a delicate problem. Successful planning requires familiarity with different methodologies and a critical understanding of them.

Developing a project plan is a one-time enterprise, but plans are by their nature capable of repeated execution. This is even true of plans which were devised to be performed once only. Project plans may be carried out in many different ways, just as computer programs have different executions. We call *plan execution* any actual realization of an appointed plan.

Throughout this book we assume plan executions to be *controlled* by an execution *supervisor* who sets activities going and makes decisions in choice situations.

Alternative plan executions will in general have a different "degree of desirability", depending on durations, costs, failure rates, setup times, etc. Choosing an optimal course of action when the plan allows for different moves is a crucial problem of plan execution supervision. It consists in working out *optimal* or at least *satisfactory plan executions* and putting them into effect. As we will see, making decisions about execution alternatives requires taking several *attributes* of operations into account — attributes often just known in "vague" form.

The first chapter of the book offers an informal survey of the methodologies presented further on. We hope that it will provide the reader both with an easy start in project engineering, and with a unitary view of our subject matter.

1. Survey of Methodologies

This chapter is dedicated to a preliminary view of project design and analysis methodologies which we will formally introduce later on. The reader should consider it a short guided tour of the different approaches, carried out with aim of giving an informal, but comprehensive, presentation of the various points of view. After that tour the reader should have a first, rough, idea of how to chose a design methodology, and how to use the rest of this book.

Every design methodology is grounded in a particular planning language and, hence, offers different analysis possibilities. Two factors are decisive for choosing the design methodology and, consequently, the planning language: available information and required outcomes. The latter embody the interests which gave rise to the project.

There is a single thread on which this informal survey is strung. It is a toy project, which we are going to design and analyze repeatedly, each time committing ourselves to a slightly different project specification. In other words: each time starting with somewhat different information and being driven by different interests.

Our specification will always explicitly state the intended use of the project plan to be constructed. We will single out this particular information by means of a black dot.

For each specification, we will indicate which planning language is best suited and why; for each planning language, we will illustrate the available analysis methods and their outcomes.

As explained in the introduction, a *project* is a piece of planned work driven by a definite interest. Our toy project is to test the prototype June_89 of a proposed product in order to enable a go/no-go decision on further product development based on June_89.

We will start our introductory tour with the examination of the following project specification.

<u>Specification 1.1:</u>

1. The prototype June_89 will undergo two tests — a performance test P and a budgetary test B. The results of these tests are necessary input to the go/no-go decision.2. Test P and test B will be carried out independently, in two different locations, LP and LB. Carrying out the tests will take, respectively, t_P and t_B time units.

3. At a certain point in time the necessary documentation will be distributed to the two testing locations. The transmission to locations LP and LB will take, respectively, s_P and s_B time units.

4. The outcomes of both tests will be examined by the project management team in order to reach the go/no-go decision. This examination will take e_P and e_B time units, in the order mentioned.

•   We need a time scheduling plan for the activities described in the procedure above, and information about the possible completion date.

This specification is very simple indeed, but nevertheless — as we will now see — sufficiently characterized to suggest using the planning language of *activity networks* .

Methodologies based on this planning language conceive a project as a finite collection of elementary activities requiring a certain time to be completed, and partially ordered by a precedence relation; the only considered events are starts and terminations of activities.

We call an activity network directed graph — more briefly expressed, a digraph — representing a project so that: arcs stand for an elementary activities; the initial and the terminal vertex of an arc representing activity a correspond, respectively, to the events 'activity a begins' and 'activity a terminates'; to each arc there is associated a real positive number which has to be interpreted as the duration of the corresponding activity.

The digraph of an activity network has to satisfy particular properties we will illustrate in Chapter 3. Here we just want to point out that activity networks do not allow the representation of alternatives or choices.

In the activity network interpretation of a digraph, vertices are synchronization points for starts and terminations of activities, and their common 'logic' is the so-called *AND/AND realization logic* : a vertex is 'realized' at the time point when all its input activities have terminated, and at that time point all its output activities will start. As a consequence, there are no means for representing the choice between starting either of two activities, nor for expressing that a certain activity will start just at the termination of either activity out of two.

But let us step back to our toy project. We notice some relevant features of Specification 1.1:

- Specification 1.1 describes a project as a finite collection of elementary activities which require a certain time to be completed and which are partially ordered by a precedence relation based on time. For example, sending a copy of June_89 to location LP precedes carrying out test P, while the executions of tests P and B are time independent.

- An event takes place when and only when all activities it terminates have been accomplished and its taking place starts all the activities which it begins. In other words, AND/AND synchronization of activities is required: documentation must be sent to the testing locations LP and LB at the same time point (AND starting); the decision over approval/rejection needs examination of the outcome of both tests (AND termination).

- Neither alternative ways to achieve subgoals nor choices of any other sort are described.

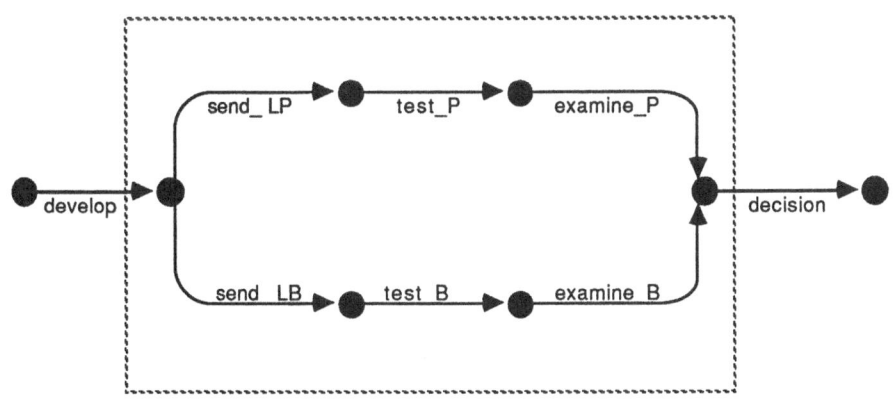

Fig. 1.1

A specification with these peculiarities leads to the planning language of activity networks, and suggests, for design and analysis, methodologies based on that language:

CPM if we are able to assign a duration to each involved activity, PERT if it appears reasonable to assume the durations of activities to be beta distributed random variables, and if we are able to estimate, for each of them, the minimum, the most likely and the maximum value.

The activity network in Fig.1.1 represents a plan matching Specification 1.1. Such a plan is not unique: other, different, plans can be designed in accordance to that specification, even other activity network plans. Fig.1.2 shows one of them.

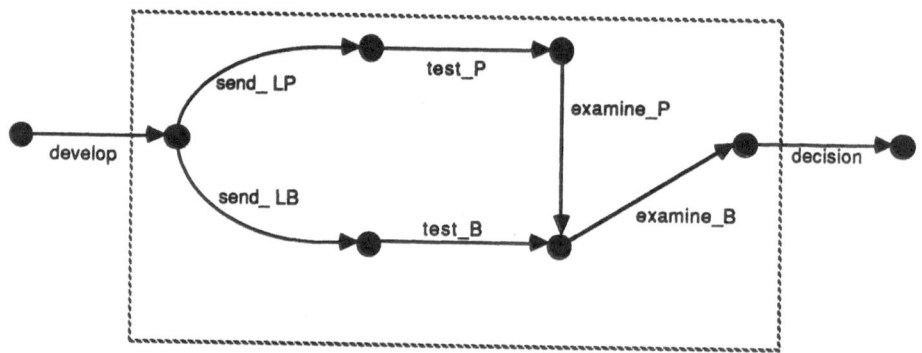

Fig. 1.2

Designing a project by means of an ordinary activity network is in general possible only if the specification set meets three requirements:
1. A partial ordering among the mentioned activities is given, expressed as precedence in time.
2. Synchronizations of activities are specified in a manner consistent with AND/AND realization logic.
3. Neither alternatives nor choices are required.

Requirements 2 and 3 typically apply to projects aimed at a unique, although possibly time dependent, even uncertain, execution strategy. Structurally different execution strategies cannot be represented in the frame of a single activity network. So, for instance, activity networks may well be used to represent a rigid assembly line, but can hardly be used to design a flexible manufacturing system. We will see further on that modified activity networks have been introduced in order to overcome these limitations.

The well-known project design techniques CPM (Critical Path Method) and PERT (Program Evaluation and Review Technique) both use the planning language of activity networks: CPM and PERT plans are acyclic activity networks with exactly one source and one sink. Formal definitions and proofs are given in Chapter 4; here we are only concerned with the usage difference between the two techniques.

This difference results from the way activity durations are considered: CPM assumes activity durations to be known non-negative reals, while PERT defines them as non-negative beta distributed random variables. As a consequence, a necessary premise for the application of the CPM technique is that the planner is able to assign a single duration value to each activity, while the use of PERT requires the estimation of three values: shortest duration, most likely duration, longest duration.

Now we will apply both techniques to our toy project to contrast their application scope.

Figure 1.3a displays the activity network of Fig.1.1 with simpler labeling and without first and last arc - cut out because the corresponding activities were not properly described in Specification 1.1.

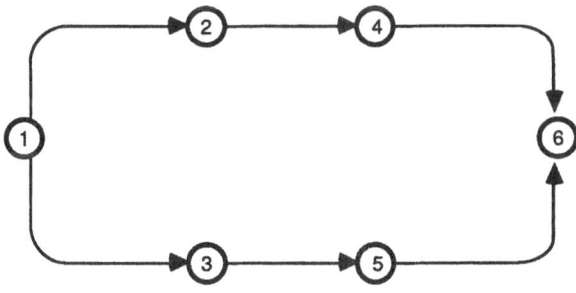

Fig.1.3a

Suppose we have, for the durations of activities, the values listed in Table 1.1. An activity network plan of this kind — acyclic, with exactly one source and one sink and where all durations are known non-negative reals — suggests the use of CPM.

CPM permits finding out the temporal constraints relative to realization of events, and the critical activities — those activities whose delay would defer the project completion date.

Table 1.1

activity	duration (time units)
1-2	2
1-3	4
2-4	15
3-5	10
4-6	1
5-6	2

Typical CPM temporal event constraints are:
- the earliest event time (ET),
- the latest event time (LT),
- the event slack (S).

The earliest event time is the earliest date an event in the network can occur; the latest event time is the latest date at which an event can take place without delaying project completion. The event slack is the difference — computed for each event — between the latest event time and the earliest event time.

Table 1.2

event	ET	LT	S
1	0	0	0
2	2	2	0
3	4	6	2
4	17	17	0
5	14	16	2
6	18	18	0

Critical activities make up the *critical path*, the longest-time path from source to sink. The duration of the critical path coincides with the minimum time which is necessary for project completion.

The critical path can be detected quite easily, since it traces through events whose slack is equal to zero. It is a relevant element for the management of project realization: critical activities are crucial to the project completion date and have to be strictly controlled and best supplied with resources.

Returning to our toy project: the values of the temporal event constraints are those listed in Table 1.2. The corresponding CPM network is the one depicted in Fig.1.3b, where the activity durations are written on the arcs and the critical path is drawn bold. It is easy to see that the minimum project completion time is 18 days.

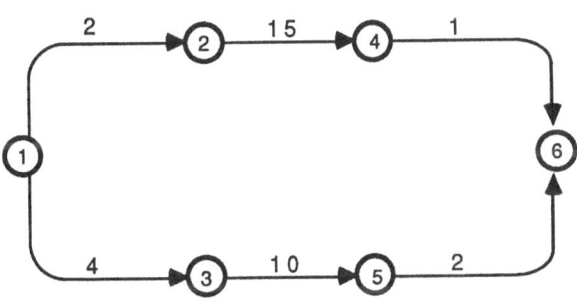

Fig.1.3b

If the activity scheduling plan is detailed enough to associate costs with activities, and if these costs are can be assumed fixed and known — remember that CPM requires each activity to have a fixed and known duration — a cost plan may be easily developed on the basis of the activity network.

The CPM cost technique will then support cost plan analysis and cost control, and in particular, the evaluation of additional financial and/or organizational costs arising when activities must be speeded up during execution.

The assumptions for applying the CPM technique — being able to assign both fixed durations and costs to activities — may not be fulfilled if durations or costs of activities exhibit a sufficient degree of uncertainty.

We shall, here, concentrate on durations. In many cases, it will be reasonable to assume the durations to be independent, beta distributed random variables for which we are able to give three time estimates: the shortest duration s, the most likely duration m, the longest duration l.

This is suitable situation for turning to the PERT technique. PERT takes uncertainty of durations explicitly into account, and, as we will see, yields easily useful results. However, the computations are facile only as long as one is ready to accept that the parameters characterizing the beta functions undergo severe limitations. This will result in a rather drastic constraining of the allowed shape: skewness is limited, and the function becomes rather flattened.

We will see in Chapter 3 that without these restrictions the application of PERT leads to cumbersome calculations.

So, we shall assume that each activity duration of our small project is a beta distributed random variable, and that we are able to provide estimates for the shortest, the most likely and longest duration of each activity. Let such estimates be those listed in Table 1.3 together with the approximated values of both mean and standard deviation.

Table 1.3

activity	s	m	l	mean	standard deviation
1-2	1	2	3	1.00	0.33
1-3	2	4	5	3.83	0.50
2-4	12	15	16	14.66	0.66
3-5	9	10	15	10.66	1.00
4-6	1	1	2	1.16	0.16
5-6	1	2	3	1.00	0.66

The computation of expected event times repeats that of CPM, except for using mean durations instead of fixed ones, and for getting expected values instead of deterministic results.

It is easy to see that the earliest expected event time for event 6 is of 17.82 time units, since the expected value for the latest completion of both activities 4-6 and 5-6 is 17.82.

The critical path, too, is defined on the basis of the mean duration of activities and leads to the minimum expected time for project completion. This calculation requires that the critical path is made up of a large number of activities, and ignores the dispersion of activity durations around their expected value.

The application of PERT should in any case follow a careful analysis of the project specification. As we will see, if the network is made up of many different source-sink paths, the source-sink path with the greatest probability to be critical may have a very — too! — low probability. Or even: non-critical paths may be more likely to violate the established project completion date than the critical path.

We shall now consider a slightly different specification for our project.

Specification 1.2:

1. The prototype June_89 will undergo two tests — a performance test P and a budgetary test B. The results of these tests are necessary input to the go/no-go decision.

2. Test P and test B will be carried out independently, in two different locations, LP and LB. Carrying out the tests will take, respectively, t_P and t_B time units.

3. At a certain point in time the necessary documentation will be distributed to the two testing locations. The transmission to locations LP and LB will take, respectively, s_P and s_B time units.

4. Both tests have two distinct, mutually exclusive outcomes, labelled 'positive' and 'negative': test P leads to outcome 'positive' with probability 0.7 while test B leads to outcome 'positive' with probability 0.8.

5. The outcomes of both tests will be transmitted to the project management team. Transmission from location LP will take p_P or n_P time units, depending on whether the outcome is positive or negative. Transmission from location LB will take p_B or n_B time units, again depending on whether the outcome is positive or negative.

6. The project management team will decide to go ahead if both test P and B yield the outcome 'positive', and will otherwise decide to stop June_89. A go decision will start the implementation of June_89, while a no-go decision will result either in its revision or rejection.

• We need a time scheduling plan for the activities described in the former procedure, and information about the decision date.

Specification 1.2 differs from Specification 1.1 in two significant features. First, the project described by Specification 1.2 has two ways to end: the go decision and the no-go decision; the project defined by Specification 1.1 has one ending: the go/no-go decision. Secondly, Specification 1.2 describes two new kinds of events involving choice:

- events which start exactly one of two next activities. Events of this sort — like both ending events of test P and of test B — are said to have EXOR (Exclusive Or) output.
- events which are specified to take place when at least one of the set of just prior activities terminates. Events of this kind — like the event which starts the no-go decision making — are said to have OR input.

In substance, Specification 1.2 describes a plan with two endings — commonly called sinks — and three sorts of event 'logic': the AND/AND logic we have already met in CPM-PERT networks, and the AND/EXOR and OR/AND logics.

Figure 1.4 shows a plan for the project defined by Specification 1.2. The representation is a digraph with three kinds of vertices, graphically distinguished by means of different shadings.

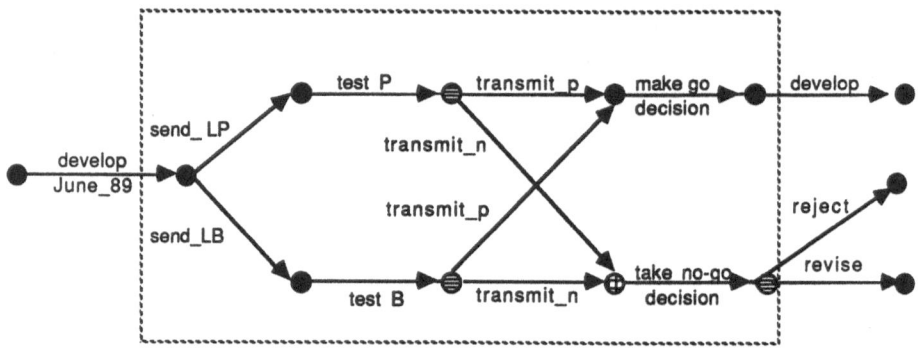

Fig. 1.4

This graph gives a preliminary idea of the planning language which goes with the analysis technique called GERT. This methodology can be used when the project specification presents the features we just illustrated — more sinks and different 'event logics'. GERT also admits project definitions involving cycles of activities.

The plan of Fig.1.4 is re-represented in Fig.1.5 by means of the GERT graphical language. The numbers inscribed in the right-hand side of vertices are simply node identifiers.

Vertex 1 is the source node, while vertices 8 and 9 are two sinks; vertices 2 and 3 are AND/AND nodes — like CPM-PERT vertices.

Vertices 3 and 5 are nodes with probabilistic EXOR output: nodes starting exactly one output activity, picked out in accordance with a given probability distribution. The probability of each output activity to be started is inscribed at the corresponding arc.

Vertices 6 and 7 represent events with OR input. These events are defined to occur as soon as some specified number n of the input activities terminate. The integer n is inscribed in the upper left-hand side of the node's graphical representation.

In particular: the vertex 6 event takes place when both its input activities are terminated (n=2); it works as an AND/AND node. The vertex 7 event takes place when at least one input activity is terminated (n=1); it is hence an OR/AND node.

GERT networks are also called generalized activity networks and stochastic activity networks.

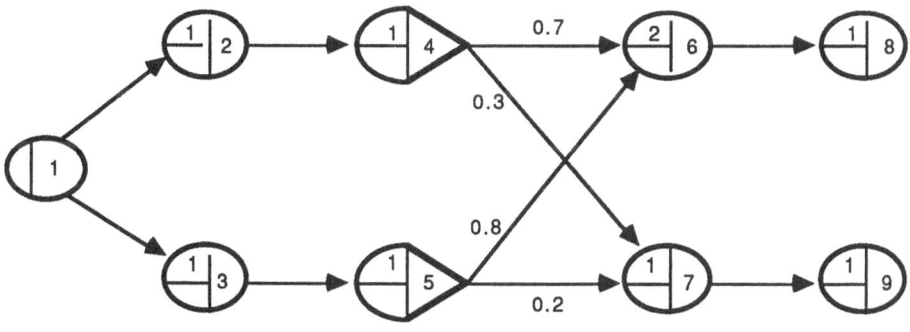

Fig. 1.5

This very small example shows that the GERT technique offers a more expressive and flexible planning language than CPM-PERT. In fact, GERT diagrams are enriched activity networks, so that every CPM-PERT plan may be seen as a particular GERT plan.

The price paid for dropping CPM-PERT limitations is renouncing symbolic analysis of the network. Symbolic analysis algorithms do exist for two particular classes of GERT

networks — STEOR and Basic Element Structures networks — but when systems of realistic size are to be considered the computational complexity of those methods suggests
resorting to simulation techniques — or, at least, to some combination of symbolic algorithms and simulation. Sophisticated simulation packages can provide time, cost and resource allocation statistics.

In summary: Specification 1.2 describes a project as a finite collection of activities
requiring time to be completed and partially ordered by a precedence relation. But the assumptions required for applying CPM-PERT are not satisfied: the activities starting at the
termination of test P — i.e., transmission of either a positive or a negative outcome from
LP — are neither both carried out nor carried out with certainty.

The same holds true for both activities starting at the termination of test B: no-go decision making starts when one of its input activities has terminated, while go decision
making requires that both its input activities have terminated.

This sort of specification suggests the GERT technique for developing a project plan.
If the resulting GERT network is a general one, analysis will have to be done by simulation.

We now consider yet another variation in specifying our toy project.

Specification 1.3:

1. The prototype June_89 will undergo two tests — a performance test P and a bud-
    getary test B. The results of these tests are necessary input to a go/no-go decision.
2. Test P and test B will be carried out independently.
3. Both tests have two distinct, mutually exclusive outcomes, labelled 'positive' and
    'negative'.
4. The project management team will decide to go ahead with June_89 if both tests P
    and B yield the outcome 'positive', and will otherwise decide to stop it.
•   We need the plan to monitor state changes in the decision procedure above and to an-
    alyze its causal structure.

The planning languages introduced above are not well suited for monitoring process
state changes — that is, for representing plan executions in progress. Since Specification
1.3 declares monitoring to be a requirement, we have to look for a different planning language allowing the representation of both project states and state changes.

Petri net theory offers a type of planning language with several variations, suited for
monitoring processes. As we will see, the Petri net variants diverge in the way disposable

information is represented. In this chapter we shall informally introduce the fundamental Petri net variants: condition/event, place/transition and predicate/transition nets. Once more, our toy project will be helpful.

The *condition/event net (CE net)* in Fig.1.6 represents a plan for Specification 1.3.

Circles represent atomic *conditions* which may or may not be fulfilled. A maximal set of conditions holding together is called a *case* in Petri net terminology, and represents a possible state of planned execution. Fulfilled conditions are graphically distinguished by a black dot — a *token* — drawn in the corresponding circle.

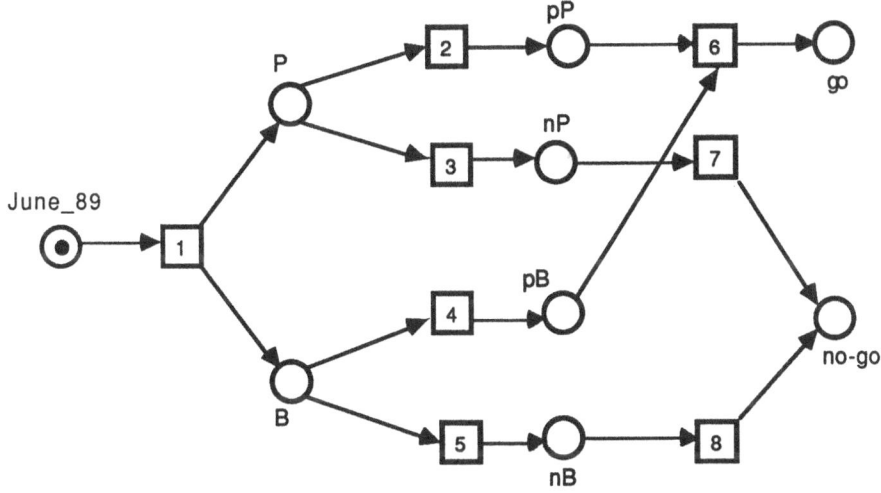

Fig. 1.6

The plan in Fig.1.6 is represented together with its initial case. The only holding condition is June_89, to be interpreted as 'Prototype June_89 is ready to be tested'.

Boxes represent *events* which may or may not take place. The taking place of events represents the changing of the plan execution state.

Directed arcs represent the causal structure of the project by associating conditions to events and events to conditions: an arc from a circle to a box indicates that the condition is a *pre-condition* of the event; conversely, an arc from a box to a circle indicates that the condition is a *post-condition* of the event.

When applying Petri nets for project planning, it is helpful to view conditions as re-source availability constraints, and events as activities, where resources and activities are to be understood in a very broad sense. We will often rely on this interpretation.

In the condition/event net of Fig. 1.6:
- circle p represents availability of the prototype June_89 for dispatch;
- box 1 represents submitting June_89 to both test P and B;
- circles P and B represent the availability of June_89 for test P and B, respectively;
- boxes 2, 3, 4, 5 represent, respectively, the events: test P yields outcome 'positive'; test P yields outcome 'negative'; test B yields outcome 'positive'; test B yields out-come 'negative';
- circles pP, nP, pB, nB represent, respectively, the availability of outcome 'positive' from test P, of outcome 'negative' from test P, of outcome 'positive' from test B, of outcome 'negative' from test B;
- box 6 represents the acknowledgement of June_89, while boxes 7 and 8 represent two different non-acknowledgement events;
- circle go represents the go decision, circle no-go the no-go decision
- p is the only condition which holds initially.

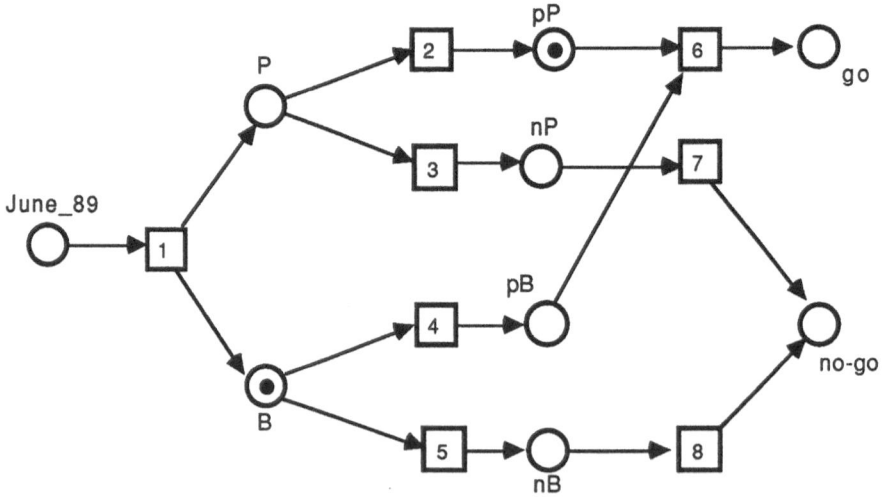

Fig. 1.7

Figure 1.7 shows the representation of another case of our toy project: outcome 'positive' has already come from test P, and is actually available, while test B is still going on.

Petri nets are a language suitable for describing the behavior of systems. Indeed, Petri nets allow the representation and monitoring of state changes based on the causal relationship between conditions and events. The engine of this capability is provided by the so-called *transition rule*:

- an event is *enabled* if all its pre-conditions are satisfied while all its post-conditions are not;
- an enabled event may (!) take place, or occur;
- after its occurrence all its post-conditions will be satisfied while all its pre-conditions will no longer be.

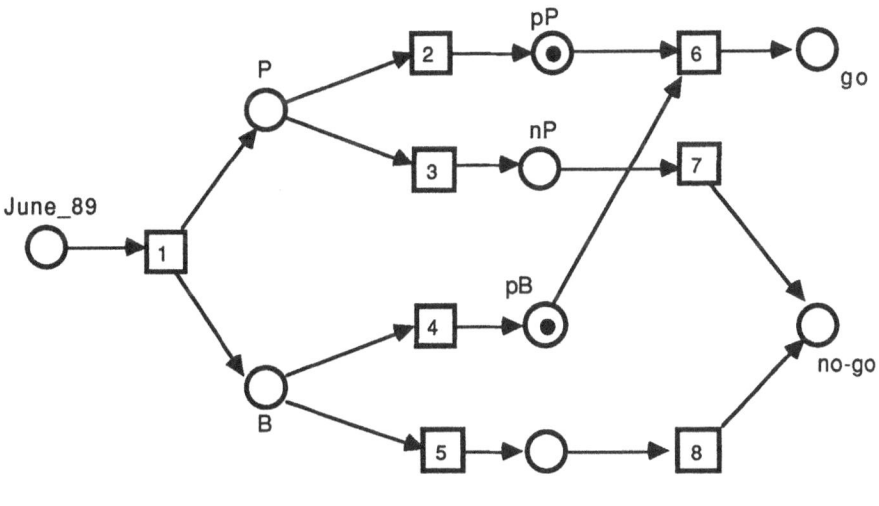

Fig. 1.8

Graphically, an event is enabled if all input circles of the corresponding box carry a token, while all output circles of the same box do not.

It is clear that the occurrence of one or more events changes the holding case. In Fig. 1.7 the enabled events are 4 and 5; event 6 is not enabled since condition pB is not satisfied. If, for instance, event 4 occurs — i.e., if B generates the outcome 'positive' — the plan will enter another case, the one depicted in Fig.1.8.

At this case the only enabled event is event 6, and after its occurrence the only holding condition will be condition go: 'the go decision has been taken'.

The simultaneous enabling of both events 4 and 5 at the case in Fig. 1.7 is such that the occurrence of either event disables the other one: after the occurrence of either event 4 or 5 both events will no longer be enabled.

A situation of this sort is called *conflict*. In a project, the crucial conflicts are those regarding the allocation of resources and those representing choices of execution strategies. The conflict in Fig. 1.7 is not of this sort; it is simply a logical conflict.

One useful feature of CE plans is the easy identification of conflicts in reachable cases.

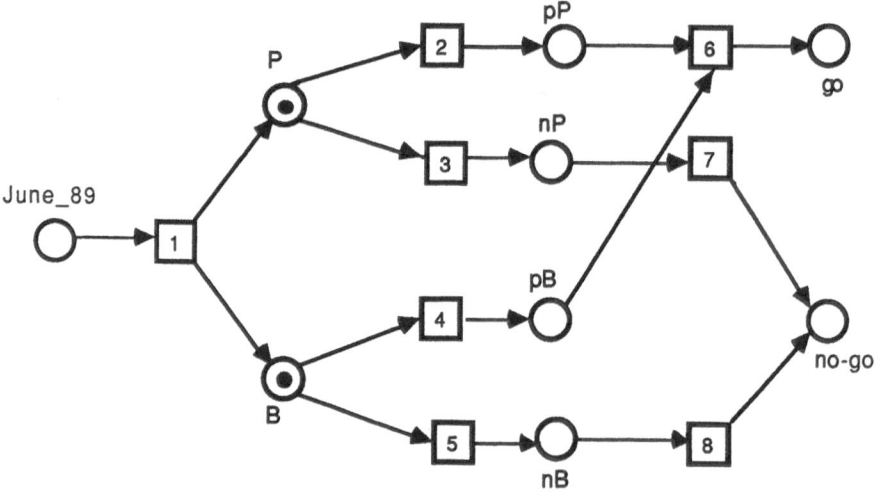

Fig. 1.9

Suppose the case depicted in Fig. 1.9 holds: events 2, 3, 4 and 5 are enabled. Events 2 and 3 are in conflict with each other, and so are events 4 and 5. But events 2 and 4, which are also both simultaneously enabled, are not in conflict, since each of them may occur without affecting in any way the enabling of the other. Their occurrence is not ordered: they can actually occur in any order.

We express this by saying that events 2 and 4 are concurrently enabled at the considered case, or, that 2 and 4 are *concurrent events*. Of course, events 2 and 5 are concurrent too, and so are events 3 and 4, and events 3 and 5.

The two conflicts — the one between events 2 and 3, and the one between events 4 and 5 — correspond to the requirement of mutual exclusion between the outcomes of a single test, while the four pairs of concurrent events listed just above express the reciprocal independence of tests P and B.

Concurrency and conflict are central concepts in the analysis of independent processes. As we will see, the theory of Petri nets provides a solid foundation for both these notions. For this reason, Petri nets recommend themselves for project development each time the interdependence of processes plays a relevant role.

In condition/event nets a circle carries one bit of information: the corresponding condition is or is not satisfied. This bit of information is graphically represented by the presence/absence of a token in the corresponding circle.

For this reason, the size of CE nets quickly becomes unmanageable when they are applied to plans in which the number of consumed or produced resources must be taken into account. In CE nets, each plan-relevant resource unit has to be represented by means of a circle. Representing thirty resource units produced, or consumed, one by one requires thirty circles.

The problem of dealing with multiple resources is neatly addressed by another class of nets, called *place/transition nets (PT nets)*. Here, the representation of multiple resources is made quite natural: circles represent "places" where several units of resources may be stored. T-elements are interpreted as elementary operations connected via resources, and the oriented arcs express consumption/production of resource units.

Consider again our toy project, now described by Specification 1.4.

Specification 1.4:

1. Ten copies of prototype June_89 will be tested: all of them will undergo the performance test P, but only three will undergo the budgetary test B. The results of the tests are necessary input to a go/no-go decision.
2. Tests P and B will be carried out independently.
3. Both tests have two distinct, mutually exclusive outcomes, labelled 'positive' and 'negative'.

4. The project management team will decide to go ahead with June_89 if both test P yielded at least eight 'positive' outcomes and test B exactly three. Otherwise, project June_89 will be stopped.

•   We need the plan to monitor state changes in the decision procedure above and to analyze its causal structure.

The *place/transition net (PT net)* of Fig.1.10 represents a plan for this project.

Circles (and ellipses) represent *places* where several *units of a specific resource may be stored*. The number of actually available units of a certain resource is called the *marking of the place*. Place markings may be directly written in the circles, or represented by sets of black dots (tokens). The distribution of resource units over the net at a given time, is called *a marking of the net*.

The plan in Fig.1.10 is represented together with its initial marking: eleven copies of the resource "June_89" are available.

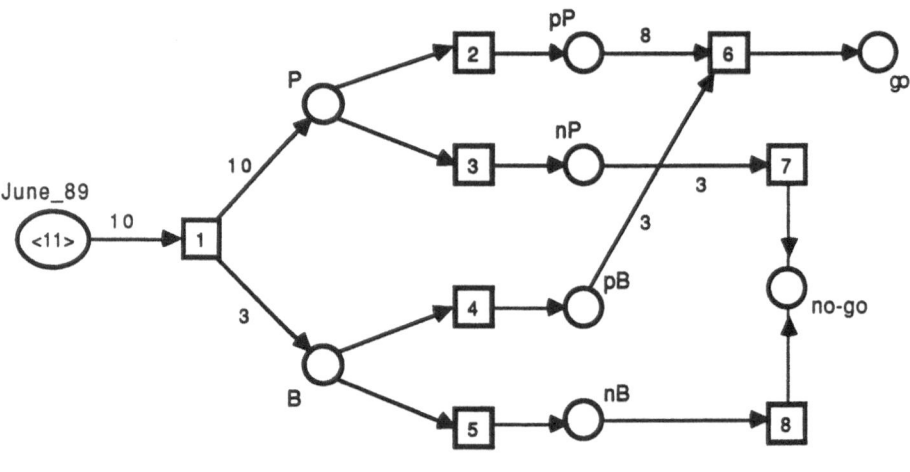

Fig. 1.10

Boxes represent *operations* which consume and/or produce resource units. A directed arc from circle A to box B represents the consumption of units of resource A by operation B. Conversely, an arc from box B to circle A indicates the production of units of resource A by operation B. The number of consumed/produced resource units is represented by the arc weight. Arc weights equal to one are usually omitted.

The interpretation of the elements of place/transition net 1.10 is quite close to that of the elements of condition/event net 1.6, so that we can leave it, as a little exercise, to the reader.

As we shall see, the analysis possibilities offered by condition/event and place/transition nets are similar. However, some unitary basic concepts, in the context of condition/event nets, turn into concepts with multiple versions in the context of place/transition nets. For instance, this happens to the notions of conflict and of concurrent occurrence of events. Therefore, condition/event nets should be used for the causal analysis of plans whenever possible, while place/transition nets should be used for modeling the flow of resources.

There are two cornerstones of Petri net analysis — and, particularly, of condition/event and place/transition nets — the reachability graph and the incidence matrix. The *reachability graph* is a rooted digraph whose nodes represent the markings which are reachable from a given initial state. The arcs represent events, or transition occurrences.

Owing to its combinatorial nature, the size of the reachability graph grows exponentially with net size, and is not useful for large Petri nets. In any case, construction and analysis of the reachability graph must be carried out by computer for nets of realistic size, and its utility therefore depends on both speed and capacity of the available computation system.

When the net size and the available computation tool allow the construction of the reachability graph, its analysis permits an easy detection of conflicts, deadlocks and loops, the answering of liveness questions, and the identification of reachable states and of paths between them, and the investigation of causal dependencies between processes. We will introduce these analysis techniques in Chapter 5.

The reachability graph of the CE net in Fig. 1.6 is given in Fig. 1.11. Cases are represented by vectors whose components correspond to conditions: a 1-entry will indicate that the corresponding condition holds, a 0-entry that it does not. All vectors in Fig.1.11 are indexed according to the following ordering of conditions:

[June_89, P, B, pP, nP, pB, nB, go, no-go].

For instance, [011000000] represents the case depicted in Fig. 1.9.

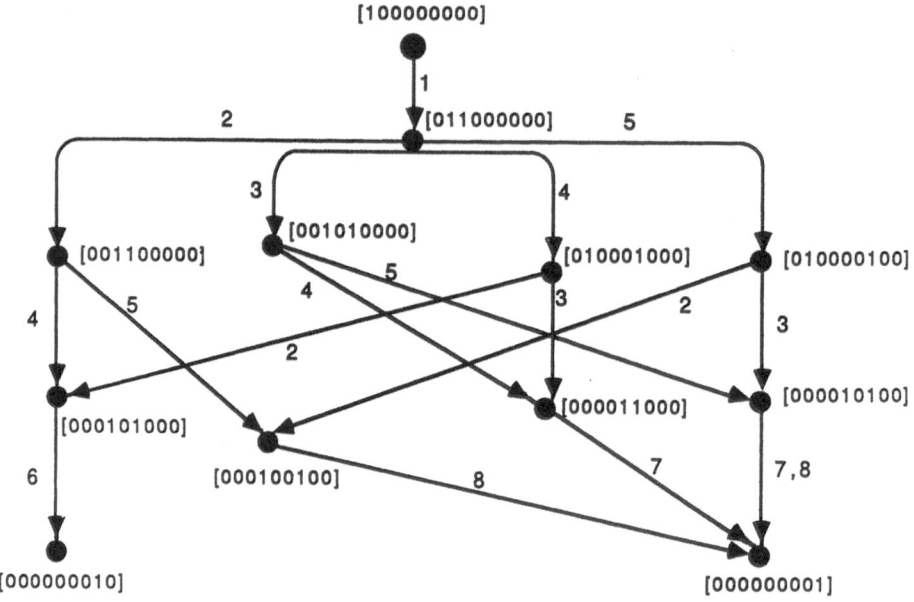

Fig. 1.11

The analysis of the reachability graph in Fig. 1.11 confirms our previous observations, and constitutes a verification of Specification 1.3:

-   case [011000000] presents two conflicts: events 2 and 3 are in conflict, and so are events 4 and 5. Conflicts are easily detected by means of the reachability graph: two events are in conflict at a certain case if they label output arcs of the corresponding node but do not belong both to the same outgoing path. The two conflicts above verify the required mutual exclusion between the outcomes of each test. No other reachable case presents conflicts.

-   Case [011000000] presents four pairs of concurrently enabled events: 2 and 4, 2 and 5, 3 and 4, 3 and 5. This verifies the specification of independence between test P and B. There are no other pairs of concurrently enabled events in reachable cases. Concurrently enabled events too may be easily detected by means of the reachability graph: two events are concurrently enabled at a certain case if they label output arcs of the corresponding node, and occur in both orders in paths going out of the considered node.

- The initial case is not live: there are two deadlocks, [000000010] and [000000001]. They correspond to the two end points of the plan, and are therefore consistent with the specifications.
- There are no loops.

The *incidence matrix* is the second cornerstone of Petri net analysis. It is defined only for Petri nets with at most one arc — weighted or not — between each pair of nodes, and furnishes an algebraic representation of the net structure.

We will present incidence matrices and related analysis methods in detail in Chapter 5. These methods focus on the calculation of net *invariants,* which represent semantically meaningful plan components.

In summary: Specifications 1.3 and 1.4 declare monitoring of state-changes to be a driving interest of the plan. Some alternatives are in conflict, and therefore a choice between them must be exercised when executing the plan. Time and probability aspects are not mentioned.

These features suggest a Petri net planning language. Specification 1.3 does not describe consumption or production of more units of the same resource, while Specification 1.4 does. This reasons alone suggest applying condition/event nets in the first case, place/transition nets in the second.

In condition/event nets circles carry one bit of information: condition x is or is not satisfied. In a place/transition nets circles carry the information how many units of resource x are actually available. Units of the same resource are assumed to be indistinguishable, though countable, entities.

However, it is often the case that units of the same resource come with variable features which make a difference in the framework of the plan. Taking these features into account mostly leads to large and complex diagrams, difficult both to understand and analyze.

Predicate/transition nets (PrT nets) allow for a drastic reduction in the number of net elements whenever the consumption or production of resource units with individual features has to be planned. Since information can be moved from the net structure to the tokens, the plan representation can be kept more concise.

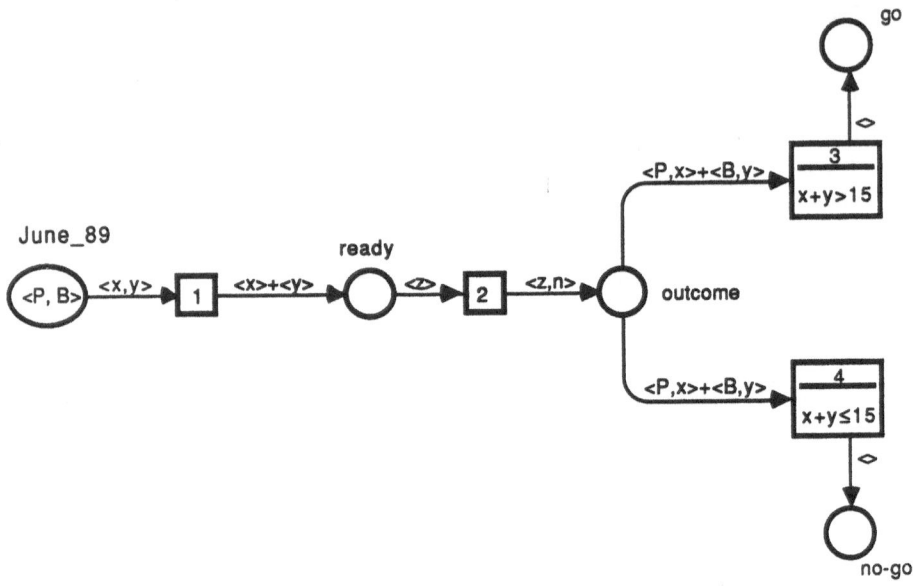

Fig. 1.12

Again, we introduce the Pr/T net representation of plans by means of our toy project. This time we assume that it is described by the following

Specification 1.5:

1. The prototype June_89 will undergo two tests — a performance test and a budgetary test. The results of these tests are necessary input to a go/no-go decision about further product development.
2. A copy of the product, P, will be tested on performance; a cost account, B, on conformity to budget.
3. The outcome of each test will be expressed as an integer value $n = 1,2, ...,10$, to be interpreted as a figure of merit, from worst to best.
4. The project management will decide to go ahead with June_89 if the sum of two figures of merit is greater than 15. Otherwise, they will stop it.
• We need the plan to monitor state changes in the decision procedure above and to analyze its causal structure.

The *predicate/transition net (PrT net)* of Fig.1.12 represents a plan for the project described as in Specification 1.5. Circles represent *predicates:*

- circle "June_89" represent the predicate "product P and cost account B are ready to be tested"
- circle "ready" represents the predicate "x ready to be tested"
- circle "outcome" represent the predicate "the test on z yielded the value n"
- circle "no-go" represents the predicate "no-go decision was made"
- circle "go" represents the predicate "go decision was made".

Boxes represent operations and are called *transitions::*
- Transition 1 represents operation "preparing the material for the tests"
- Transition 2 represents operation "carrying out a test"
- Transition 3 represents operation "making go decision"
- Transition 4 represents operation "making no-go decision".

Each predicate has a specific number of variables entries. Predicates "ready", "go" and "no-go" have one one variable entry; "June_89" and "outcome" have two.

Predicate variables range over a specific set of constants. Therefore, every n-ary predicate ranges over sets of constant n-tuples, called the *color set of the predicate.* The predicate colors of PrT net 1.12 are:

predicate	colors
June_89	<P, B>
ready	<P>, <B>
outcome	<P, 1>, <P, 2>, ... , <P, 10>
	<B, 1>, <B, 2>, ... , <B, 10>
go	<>
no-go	< >

Arcs are labelled by a multiset of tuples whose entries are in one-to-one correspondence with the variables of the adjacent predicate.

In PrT net plans, execution states are represented by associating a — possibly empty — multiset of colors with each predicate. Such a multiset is called the *marking* of the associated predicate. A transition t is *enabled for a constant tuple* k, if

-   k belongs to a given set of constant tuples, called the *color set* of transition t;
-   there exists a consistent substitution of k into all arc labels around t which gives for each input arc a multiset of constant tuples not greater than the actual marking of the adjacent predicate, and respects the color sets of both input and output predicates.

A transition enabled for a certain constant tuple k, may *occur for* k. After this occurrence the marking of the input and output predicates of the transition are changed according to the corresponding arc labels.

In Fig.1.12, transition 1 is enabled for <P,B>. After its occurrence for <P,B>, 1 is no longer enabled, while transition 2 will be enabled both for <P> and for <B>. Observe that transition 3 is not allowed to occur for <1>, <2>, ... and <15>.

It is instructive to point out the features of Specification 1.5 that suggested the use of predicate/transition nets. Like in Specifications 1.3 and 1.4, monitoring state-changes is declared to be the main use of the plan. Here too, some activities are to be carried out as alternatives to one another and time and probability aspects are not explicitly considered.

Specification 1.5 differs from the previous two in that some resource units — the units of resource "outcome" — have individual features which must be distinguished in order to execute the described plan. In fact, the figures of merit associated with "outcomes" must be summed up in order to continue executing the plan.

Using PT nets to represent the plan described by Specification 1.5 would require a place for each of the twenty different units of resource "outcome", and worse, would require a transition for every pair of integer values.

When consumption or production of resource units with individual features has to be planned, PrT nets are the suitable representation language. The price for the reduction in net size — often a very large reduction — will be the loss of analysis possibilities.

The structural features of project specifications must guide the planner in the choice of a suitable planning language, together with the intended use of the plan. In practice, many different plans will be necessary for making a project realizable: scheduling plans, cost plans, plans on resource flow, etc.

And now the question arises: is it possible to integrate heterogeneous, possibly conflicting, plan attributes for choosing an optimal course of action during the realization of a project? In general, plans allow for alternative executions, and different executions always have different times, costs, failure rates, setup times, etc. Choosing an optimal course of action is a crucial question when running plans which allow for alternative moves.

The last part of the book is dedicated to this problem. We present *Operational Decision Making* (ODM), a methodology for making operational decisions in executing a plan. ODM is intended for supporting the work of *plan execution supervisors* in making decisions about alternative execution strategies. The valuations that will govern choice can be expressed as "vague" properties of operations, in the form of *fuzzy attributes*. These valuations may as well represent beliefs of the supervisor, or they may be determined as a part of building the plan.

ODM requires plans where, by construction, alternative courses of action do not interact, and the choice between alternatives is controlled by one decision. *Control nets* are a class of place/transition nets defined with this issue in mind. In plans expressed by control nets the independence of execution alternatives is guaranteed. Decisions are only made at *control places,* special places associated with the sets of alternative executions — one control place for each such set.

We end this chapter with an informal presentation of the ODM technique. A new specification of our toy project will be our starting point.

Specification 1.6:

1. Ten copies of prototype June_89 must be tested: all of them must undergo performance test P, three of them must undergo the budgetary test B.
2. Tests P and B will be carried out independently.
3. Test P can be carried out by either of teams p1 or p2; test B by either of teams b1 or b2.
4. Team p1 is to be considered as more careful, but slower and more expensive than team p2. Team b1 is to be considered more reliable but more expensive than team b2.
•   We need a plan for controlling the execution of the testing procedure; in particular, for making an optimal choice of testing teams.

The control net plan of Fig. 1.13 represents the testing procedure described by Specification 1.6. In this net:
-   circle June_89 represents the available copies of prototype June_89. In the depicted case, fifteen copies are available.
-   Box s represents sorting out ten copies of June_89 for test P, and three for test B.
-   Circles P and B represent the prototype copies ready for test P and test B, respectively.

- Boxes p1, p2, b1 and b2 represent the operations: "team p1 carries out test P", "team p2 carries out test P", "team p1 carries out test B", "team p2 carries out test B" — in the same order.
- Circles P' and B' represent tested prototype copies — P' tested on performance and B' on budget conformity.

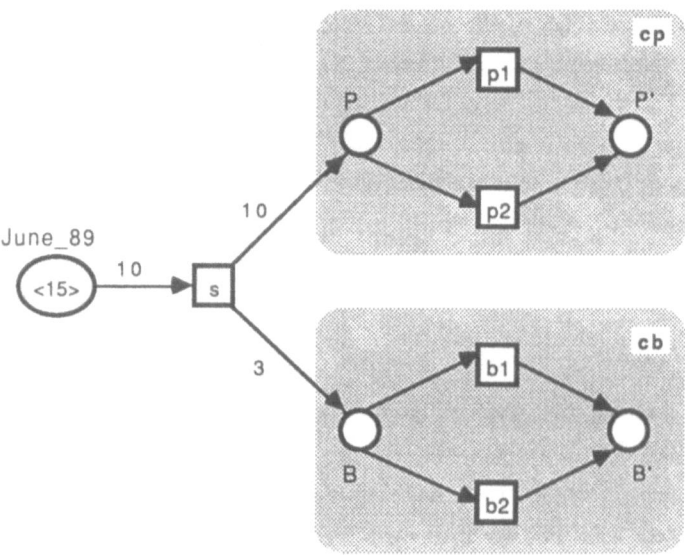

Fig. 1.13

In control net plans choice is only represented by means of *choice schemata,* special subnets made up of one entry and one exit place joined by "strands" which never interfere with one another. The entry place of such schemata is the site where decisions over the next move are made, and is therefore called the *control place.* Once the next move is made, the execution of the choice schema can proceed only along the corresponding "strand", or "strands". No further decisions will be required until the end node of the choice schema is reached. For this reason, these "strands" are called *alternatives.*

Control net 1.13 has two choice schemata, **cp** and **cb**, shadowed in the figure. **cp** represents the choice between test team p1 and p2. Its control place is place P, and it has two alternatives: transition p1 and transition p2. **cb** represents the choice between test team b1 and b2, and its control place is B. Its alternatives are transition b1 and transition b2.

ODM assumes that when the planned testing procedure is executed, an *execution supervisor* will be in charge of activating the operations involved. To do this, he will have to make decisions in choice situations. It is assumed that he will want to make "good" decisions on the basis of the available information about operation attributes and planned goals.

ODM is a multiattribute methodology for supporting decision making about alternative plan executions. It requires that the choice-relevant attributes of plans refer to the operations involved, and express desirable features. For every choice schema a set of *fuzzy attributes* must be specified by the supervisor. A fuzzy attribute A of a choice schema c is specified by assigning, for each operation t involved in c, a real number between zero and one, indicating the degree to which operation t exhibits attribute A.

A possible assignment of the fuzzy attributes mentioned in point 4 of Specification 1.6 for the choice schema whose control place is P could be:

e	careful	quick	cheap
p1	0.95	0.50	0.20
p2	0.45	0.65	0.70

Observe that attributes "slower" and "more expensive" have been substitutes by their opposites, which express desirable features of operations. The above fuzzy attributes tell us, for instance, that operation "testing by team p1" may be considered careful to degree 0.95, quick to degree 0.50, and cheap to degree 0.20.

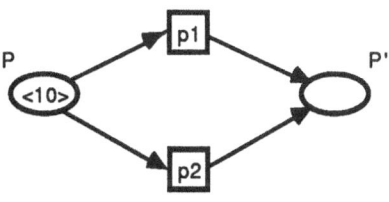

Fig. 1.14

Figure 1.14 shows an execution state of choice schema **cp** of net 1.13. Eleven different *complete* executions of **cp** are possible: carrying out operation p1 ten times, carrying out operation p1 nine times and operation p2 once, carrying out operation p1 eight times and operation p2 twice, etc. The desirability of these eleven executions will in general be different, depending on the attributes of operations p1 and p2.

Executions of choice schemata are usually made up of many operations. As we will see in Chapter 7, ODM makes it possible to extend fuzzy attributes of operations to fuzzy attributes of executions, and to associate different degrees of "importance" with each attribute. In this initial survey we are only providing a first impression of the use of ODM.

Optimal executions can be computed as intersections of execution attributes. Table 1.4 lists the fuzzy attributes over choice schema **cp**. An execution consisting of i iterations of operation p1 and $j = 10 - i$ iterations of operation p2 — carried out in any order — is denoted by "i, j". The degree to which an execution exhibits a certain attribute is computed as the weighted sum of the degrees to which the different alternatives exhibit that attribute. The weight associated with an alternative in the choice schema is the relative frequency of that alternative in the considered execution. For instance, weight 7/10 is associated with alternative p1 when computing fuzzy attributes of the execution "7, 3".

Table 1.4

execution	careful	quick	cheap
10, 0	0.95	0.50	0.20
9, 1	0.90	0.51	0.25
8, 2	0.85	0.53	0.96
7, 3	0.80	0.54	0.35
6, 4	0.75	0.56	0.40
5, 5	0.70	0.57	0.45
4, 6	0.65	0.59	0.50
3, 7	0.60	0.60	0.55
2, 8	0.45	0.62	0.60
1, 9	0.50	0.63	0.65
0,10	0.45	0.65	0 .70

Executions are then "preferable" to the degree that they exhibit all considered attributes. Table 1.5. lists the *preference degree* of the executions of choice schema **cp**.

Table 1.5

execution	preference degree
0, 0	0.20
9, 1	0.25
8, 2	0.53
7, 3	0.35
6, 4	0.40
5, 5	0.45
4, 6	0.50
3, 7	0.55
2, 8	0.45
1, 9	0.50
0,10	0.45

Most desirable executions are optimal executions. Choice schema **cp** has one optimal execution — execution "3, 7" which consists in carrying out, in any order, operation p1 three times, and operation p2 seven times.

If the preference degree of the optimal executions appears sufficient to the supervisor, he or she will put one of them into action.

ODM also includes a technique for *classifying executions* of choice schemata into indifference classes representing preference categories of the supervisor. This technique — called *fuzzy outranking* by B. Roy, its originator — is appropriate when we assume that the supervisor makes decisions with "weaker rationality". The degree to which an execution *outranks* another is computed on the basis of various special *thresholds* assigned by the supervisor. The method is too technical to be further described here, and we postpone it to Chapter 7.

ODM requires the support of computer tools. The methodology applies to graphical plan representations, which are excellent for visualization, but difficult to manipulate, especially when large. Also, to be useful, the computations must run fast enough to be part of controlling plan executions in real time.

In fact the possibility of applying computer tools is assumed for all methodologies presented in this book. In realistic cases all the methods involve the manipulation of large graphs. They are not realistic without computer aid. It is no surprise that they were all developed after computers and computer science had become part of our culture.

2. Planning with Temporal Constraints

Graphs are a mathematical model for representing systems as a relation between elementary components. The conceptual and computational tools offered by this formalism made graphs a widely applied methodology for the description and analysis of systems.

All techniques presented in this part of the book are based on representation languages which ultimately are enriched graphs - graphs to which some sort of information has been added. Therefore we devoted the following section to graph theoretical basics.

Activity networks are then introduced within the framework of the application of graphs to the representation of plans.

The introduction and demonstration of some general principles for the disciplined development of plans will conclude this chapter.

2.1 Graphs, Weighted Graphs, Bipartite Graphs

We call *graph* a pair $G = (V, E)$ where V is a finite non-empty set of elements called *vertices* and E is finite collection of pairs { v, w } of vertices.

The elements of E are called *edges* ; E and V the *edge set* and *vertex set* of **G**, respectively. The pairs belonging to E are not ordered, so that { v, w } = { w, v }. We will write vw for the edge { v, w }.

Edges are not assumed to be distinct: more than one edge vw is allowed. Collection E is in fact a *multiset* , a set mapped into the set of non-negative integers.

If e = vw is an edge of **G,** we say that v and w are the *ends* of e, that v and w are *adjacent* , that e *joins* v and w, that **e** is *incident* both to v and w and vice versa. An edge **e** = vv will be called a *loop.*

We call the *vertex adjacency function* the function which assigns for each pair v and w of vertices the number of edges incident to both vertices.

The cardinality of **V** is called the *order* of **G.**

A *graphical representation* of a graph can be obtained by representing each vertex by a dot, and each edge by an arc between the two dots representing its ends.

Figures 2.1 and 2.2 display the graphical representation of two different graphs.

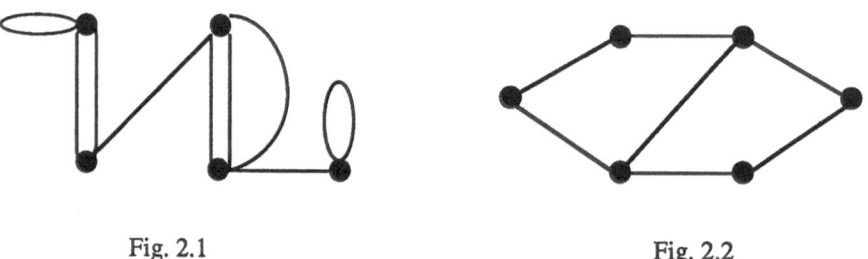

Fig. 2.1   Fig. 2.2

The number of edges incident to vertex v is the *valency* of v; a vertex of valency zero is said to be *isolated* , while a vertex of valency one is said to be an *end vertex* . We assume all graphs mentioned from now on in this book to be free of isolated vertices.

Two or more edges $e_1 = vw$, $e_2 = vw$, $e_3 = vw$ ... which join the same two vertices may be regarded as a unique *multiple edge* .

This though leads to the so-called *weighted graphs* - graphs in which an edge labelled n is used as a shorthand for n edges with the same two ends. The natural number n is called *weight of the edge* .

Sometimes, singling out a special vertex of the graph can be useful. Such a distinguished vertex will then be called the *root* of the graph, and we graphically represent it as a small square.

Figures 2.3 and 2.4 show - respectively - the graphical representation of a weighted and of a rooted graph. The weighted graph in Fig. 2.3 has been obtained from the graph of Fig. 2.1 by substituting weighted edges for multiple edges.

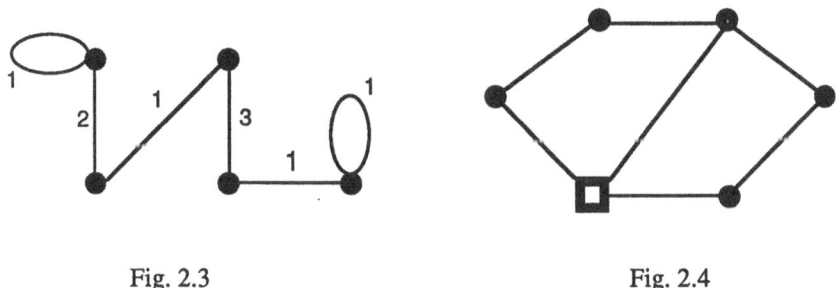

Fig. 2.3   Fig. 2.4

Two graphs G_1 and G_2 will be said to be *isomorphic* if their sets of vertices can be put in a one-to-one correspondence ϕ preserving vertex adjacency.

More formally: two graphs $G_1 = (V_1, E_1)$ and $G_2 = (V_2, E_2)$ will be said to be *isomorphic* if there exists a bijection $\phi : V_1 \rightarrow V_2$ such that
$$\forall\, a, b \in V_1 \quad \forall\, a', b' \in V_2 : \quad a' = \phi(a) \;\wedge\; b' = \phi(b) \quad \rightarrow \quad \alpha(ab) = \alpha'(a'b')$$
where α and α' are the vertex adjacency functions of G_1 and G_2, respectively.

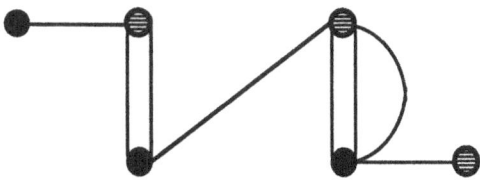

Fig. 2.5

A *bipartite graph* is a graph whose vertex set is partitioned in two subsets so that adjacent vertices never belong to the same subset.

More formally: a bipartite graph is a graph (V, E) whose vertex set is partitioned into two non-empty disjoint sets, V_1 and V_2, so that
$$\forall \, ab \; \in E : \quad a \in V_1 \; \leftrightarrow \; b \in V_2.$$
Figure 2.5 shows a bipartite graph.

Subgraphs

Let $G = (V, E)$ and $G' = (V', E')$ be two graphs. We say that G' is a *subgraph* of G if $V \supseteq V'$ and $E \supseteq E'$.

The subgraph $G' = (V', E')$ of G will be said to be a *subgraph induced by* V' in G if $E' = \{ \, vw \mid \, vw \in E \, \wedge \, v, w \in V' \, \}$.

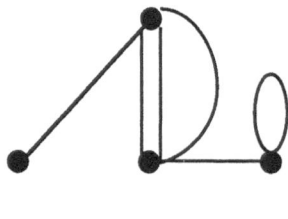   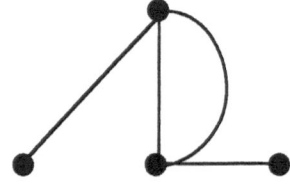

Fig. 2.6   Fig. 2.7

Both graphs in Fig. 2.6 and 2.7 are subgraphs of the graph in Fig. 2.1. Of these two only the graph in Fig. 2.6 is induced in the graph of Fig. 2.1 by its vertex set.

Digraphs

Digraphs - directed graphs - are graphs in which the edges are oriented. More formally:

We call *digraph* a graph $G = (V, A)$ if the elements of A are *ordered* pairs of vertices. The elements of A are called *arcs*. A is the *multiset of arcs*, and we will denote by $v \rightarrow w$ the arc (v, w).

The graphical representation of a digraph is the same as for an ordinary graph, except for arrows displaying arc orientation.

If multiset A contains two or more instances of the same arc v → w, those instances are called *parallel arcs* .

An arc vv is called a *loop* , and a digraph with neither loops nor parallel arcs is said to be *simple* .

The graph obtained by ignoring all arc orientation of a digraph is called its *underlying graph* .

All definitions about graphs given above immediately carry over to digraphs. So, we may speak of a *weighted digraph,* a *bipartite digraph,* a *sub-digraph,* etc. A weighted digraph is often called a *network* .

Figure 2.8 shows a digraph with both a loop and parallel arcs; Figure 2.9 a simple bipartite digraph.

Fig. 2.8   Fig. 2.9

Digraphs may be used in two different ways for representing systems as partial or full orderings of elementary components: system components may be represented either by arcs or by vertices. In this book we will often take the first point of view.

It is worth noticing here that when system components - and only system components - are represented by means of arcs, a digraph representation will not cover every conceivable ordering of system components.

For instance, consider the system Σ made up of six activities labelled 1 to 6 and ordered by following precedence constraints:

1 precedes both 3 and 4	4 precedes 7
2 precedes 5	5 precedes 7
3 precedes 6	6 precedes both 7 and 8.

If we chose - for some good reason - to represent the activities of Σ by means of arcs, there is no six-arc digraph representing the partial ordering above fully.

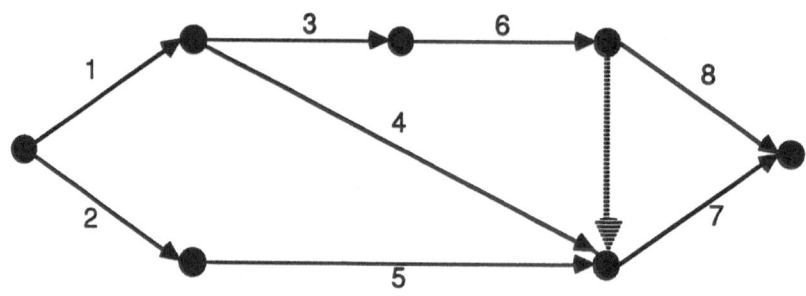

Fig. 2.10

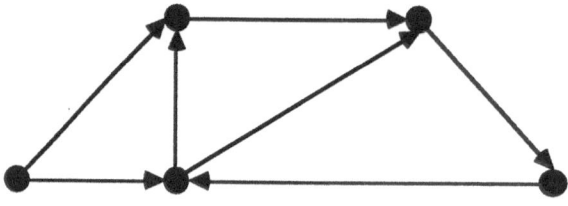

Fig. 2.11a

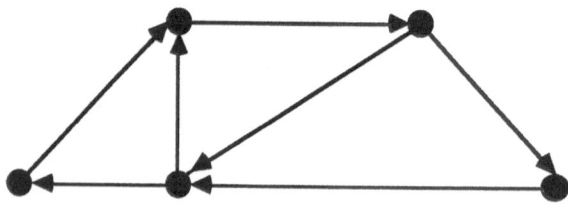

Fig. 2.11b

This difficulty may be overcome by introducing *dummy arcs* into the digraph. Figure 2.10 represents the idea for system S above: the activities are represented by arcs; a dummy arc — dotted in our figure — has been inserted in order to represent constraint '6 precedes 7'.

Walks in Graphs and Digraphs

Let $G = (V, E)$ be a graph. A *walk of length* l *from* v_1 *to* v_k is a sequence of l edges $v_1 v_2, v_2 v_3, \dots, v_{k-1} v_k$.

v_1 and v_k are called respectively the *initial* and *terminal vertex* of the walk. A walk is *closed* if $v_1 = v_k$.

A walk is called a *path* if all vertices are distinct. A closed walk in which all vertices are distinct except for the initial vertex which coincides with the terminal one, is called a *circuit*.

These concepts extend in a natural way to digraphs, where we speak of *directed walks, directed paths, directed circuits*. Directed circuits are also called *cycles*.

We say that a vertex w is *reachable* from a vertex v if there exists a directed path from v to w.

Connected Graphs, Weakly and Strongly Connected Digraphs

A *graph* is said to be *connected* if each pair of distinct vertices is joined by a path — otherwise, it is called *disconnected*.

A *digraph* is said to be *connected* if the underlaying graph is so. It is said to be *weakly connected* if for each pair of distinct vertices v and w there exists a directed path from v to w, or from w to v.

A digraph is called *strongly connected* if for each pair of distinct vertices v and w there exists both a directed path from v to w, and from w to v.

The digraph in Fig. 2.11a is weakly connected; the one in Fig. 2.11b is strongly connected. Neither graph is a tree.

Trees

A *tree* is a connected circuit-free graph; a *directed tree* is a digraph whose underlying graph is a tree. Figure 2.12 shows some trees.

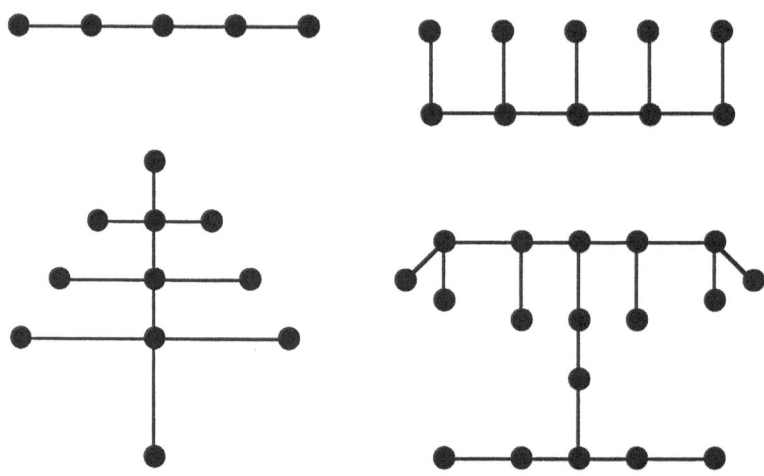

Fig. 2.12

A tree may be rooted or not. A directed rooted tree is called *out-tree* if there is a directed path from the root to any vertex. It is called *in-tree* if there is a path from any vertex to the root. In other words: in an out-tree all arcs point 'away' from the root, in an in-tree all arcs point "towards" the root.

Figure 2.13 shows three directed trees: one of them is an out-tree, one is an in-tree.

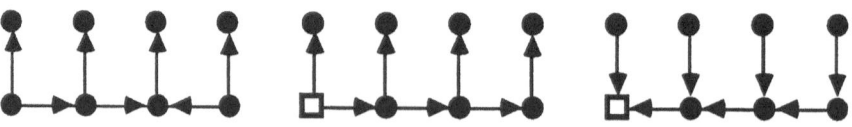

Fig. 2.13

Graph Matrices

Let $G = (V, E)$ be a graph with $V = \{v_1, v_2, \ldots, v_n\}$ and $E = \{e_1, e_2, \ldots, e_m\}$.

The structure of a graph may be represented by means of its incidence matrix, while the isomorphism between two graphs is reflected by their adjacency matrices:

the *incidence matrix* of G is the $n \times m$ matrix $C = [c_{ij}]$ where

$$c_{ij} = 1 \qquad \text{if vertex } v_i \text{ is incident to edge } e_j$$
$$c_{ij} = 0 \qquad \text{else;}$$

the *adjacency matrix* of G is the $n \times n$ matrix $A = [a_{ij}]$ where a_{ij} is the number of the edges which are incident to both v_i and v_j.

Two graphs G_1 and G_2 are *isomorphic* if and only if there exists a permutation π of vertices of G_1 such that the adjacency matrix of G_1 will coincides with that of G_2 when its rows and columns are both permuted by π.

Permutation π corresponds to renaming the vertices of G_1. If after suitable renaming of vertices, G_1 and G_2 have the same adjacency matrix, G_1 and G_2 will be considered two instances of the same graph. Unfortunately, this does not give a good algorithm for checking graph isomorphism. Running through all $n!$ permutations of vertices of G_1 in order to look for a permutation converting the adjacency matrix of G_1 into that of G_2 has exponential complexity.

Notice that graphs with the same incidence matrix are always isomorphic.

If the above G is a digraph, the existence of a directed path between two vertices is registered in the *reachability matrix* of G. This is the $n \times n$ matrix $R = [r_{ij}]$ defined by:

$$r_{ij} = 1 \qquad \text{if } v_j \text{ is reachable from } v_i$$
$$r_{ij} = 0 \qquad \text{else.}$$

Digraph Representation of Plans

In Chapter 1 we have taken a first look at a representative set of techniques for the development and analysis of project plans. We have seen that many quite different planning languages are based on the digraph representation of such plans.

Digraph representations are suitable for plans conceived as sets of activities interrelated by the circumstance that the beginning of some activities relies on the completion of others. This is the case, for instance, with production system plans, where generally the activation of some manufacturing process relies on the completion of some tasks.

Some planning languages based on digraphs simply attach a particular interpretation to vertices and edges, other provide various enrichments of the digraph definition. Activity networks are a basic planning language of the first sort, and we will introduce them directly.

2.2 Activity Networks

An *activity network* is a digraph representation of a plan, carried out under the following assumptions.

The plan is specified as a finite set A of non-overlapping non-repeatable subplans regarded as elementary, and called *activities*.

The set of activities is partially ordered by a binary transitive acyclic relation. This relation is called *precedence relation* , and denoted by the infix symbol " < ".

Completion of activities requires *time* and there is no time in which "nothing happens": at every time point at least one activity is going on. Waiting is also counted as an activity. A *finite duration* is associated with each activity, and each activity has a *beginning* and an *ending*.

Each pair (IN, OUT) of *maximal* not both empty sets of activities such that
$$\forall\, a \in IN \quad \forall\, b \in OUT: \quad a < b \quad \wedge \quad \neg\,(\exists\, c \in A: \; a < c < b)$$
is interpreted as a time point, *the* time point τ in which all activities belonging to IN are completed and all activities belonging to OUT start. Such a time point is called an *event* or *milestone*. Activities of IN end before or at time point τ, and at least one of them ends at τ; activities of OUT start exactly at time point τ.

There exist, moreover, exactly one event (Ø, OUT), called the *source*, and exactly one event (IN, Ø), called the *sink*. The source represents the beginning of the plan, the sink its completion.

It is customary to refer to the event notion above by saying that events of activity networks are *governed by AND/AND logic*. Note that, although events are defined as indivisible entities, each event has two faces - the termination of the set of input activities, and the starting of the set of output activities. Both faces have more aspects, each one relating to termination or start of a specific activity.

Formally: An *activity network* is a triple (E, A, d), where (E, A) is an out-tree with exactly one sink, and where d is a function from A into the set of non-negative reals.

The elements of vertex set E are called *events,* those of arc set A *activities.* We say that function d assigns a *duration* to each arc. Arcs with zero duration are called *dummy arcs,* or dummy activities. They do not represent actual plan activities, but express precedence constraints.

Planning with Activity Networks

The rather natural idea of representing activities by means of arcs proportional to their duration is unfortunately unfounded, because activities need not satisfy the triangle inequality. For instance, in the activity network of Fig. 2.14, there is no reason to assume that duration of activity develop_C is less than or equal to the sum of durations of activities develop_A and develop_B.

The choice of activities to specify depends on the managerial level to which the planning effort relates. More specialized plans will be devised subsequently, on the basis of more detailed specifications appropriate to lower level control.

Table 2.1

develop_A
develop_B
develop_C
develop_D
develop_E
assembly_AB
assembly_(AB)C
assembly_DE
assembly_(ABC)(DE)

The choice of activities precedes project specification, and is a design step requiring particular care. We will illustrate this point with the help of the next example.

Consider the project of developing a computer program made up of five modules: A, B, C, D and E. We first define the plan in terms of activities which refer to the modules, but not to their parts. Such activities are listed in Table 2.1.

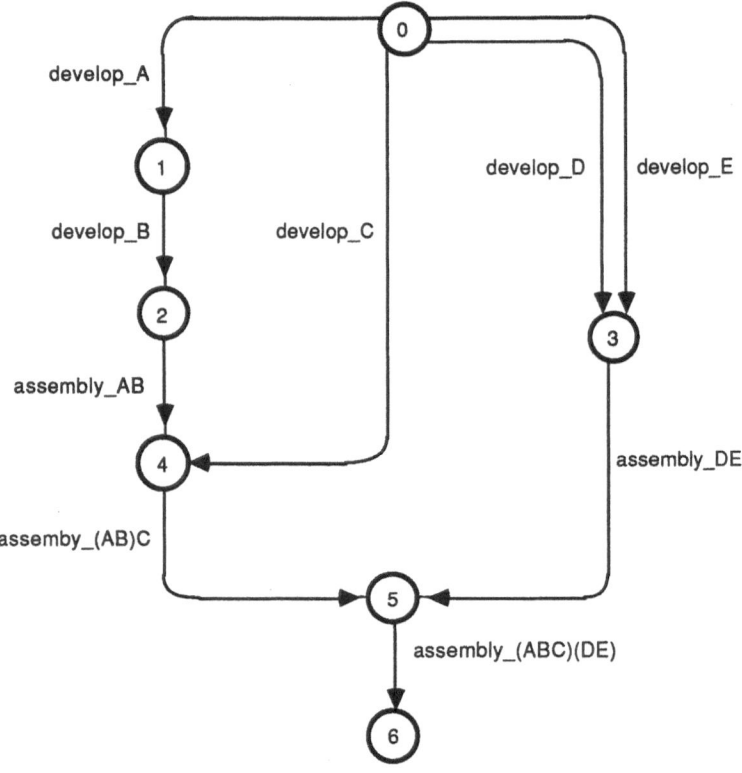

Fig. 2.14

This choice results a in project specification of specific level of detail - for instance, in Specification 2.1.

Specification 2.1:

1. Five program modules - named A, B, C, D and E - must be developed and assembled.
2. Module A must be developed before module B.
3. Module A will be assembled with module B, module D with module E.
4. Assembly AB will be put together with module C, the resulting assembly ABC with assembly DE.

An activity network plan satisfying Specification 2.1 is shown in Fig. 2.14.

Now suppose one takes the subplans listed in Table 2.2 as the elementary activities of the plan to be represented. This choice will certainly result in project specifications which are more detailed than Specification 2.1. It could - for instance - result in Specification 2.2 below.

Table 2.2

design_A	implement_A	transfer_Peter
design_B	implement_B	assembly_AB
design_C	implement_C	assembly_(AB)C
design_D	implement_D	assembly_DE
design_E	implement_E	assembly_(ABC)(DE)

Specification 2.2:

1. Five program modules - named A, B, C, D and E - must be designed, implemented and assembled.
2. The design of the five modules will be started together, and carried out independently.
3. Design of module A must terminate before starting implementation of module B, because resource 'Peter' must be transferred from activity design_A to activity implement_B.
3. After implementation, module A will be assembled with module B, module D with module E.

4.   Design of module C must be completed before assembly of modules D and E starts.
5.   Assembly AB will be put together with module C, and the resulting assembly ABC with DE.

The activity network of Fig. 2.15 represents a plan satisfying Specification 2.2.

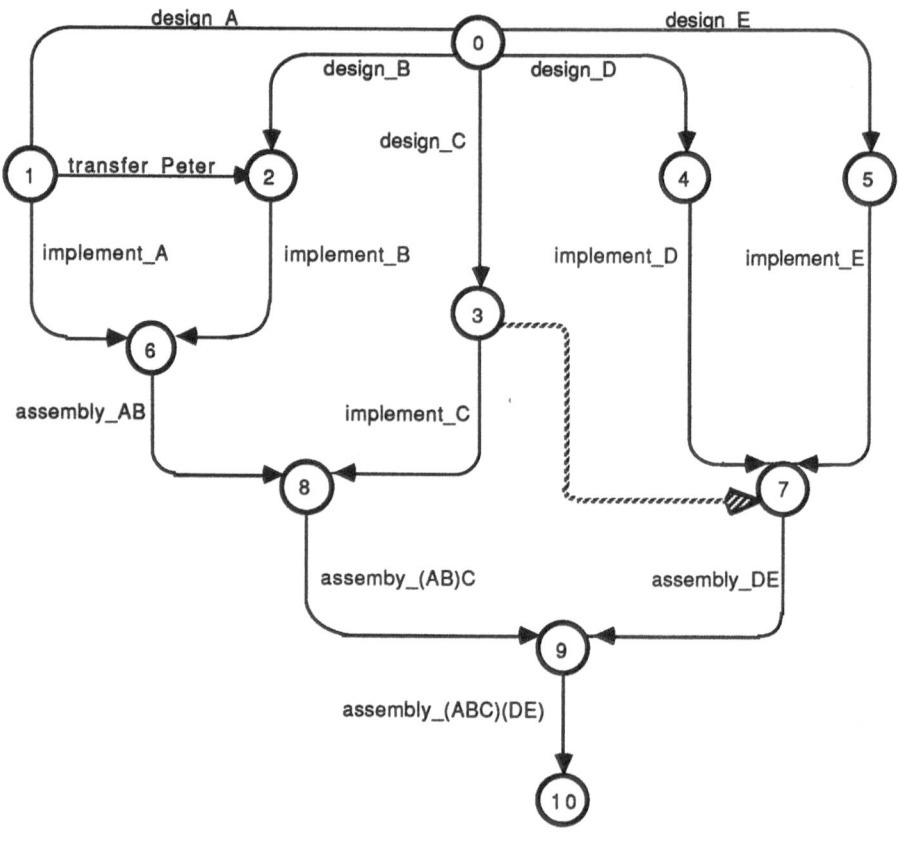

Fig. 2.15

Taking the activities of Table 2.2 for the project specification, implies a lower level of detail, and therefore leads to a quite different plan. A plan not specifiable by means of the activities in Table 2.1.

It is worth observing that, whereas the activities of Table 2.2 may be seen as obtained by subdividing the activities of Table 2.1, activity network 2.13 is not an unfolding of activity network 2.14. The following section will clarify this point.

Activity Network Unfoldings

Let $N_1 = (E_1, A_1, d_1)$ and $N_2 = (E_2, A_2, d_2)$ be two activity networks and σ a one-to-one mapping of A_1 to a proper partition of A_2.

Then we will say that A_2 is a *subdivision* of A_1, that σ is the *split function* of the subdivision, and that an *activity* $a \in A_1$ has been *split* into the set of activities $\sigma(a) \in A_2$.

We say that N_1 is an *unfolding of* N_2 if the following three conditions hold true:

(i) A_2 is a subdivision of A_1 with split function σ;

(ii) for each activity a of A_1 the directed subgraph induced by $\sigma(a)$ in A_2 has exactly one source and one sink;

(iii) if $a_1, a_2 \in A_1$ are *sequential* - that is, if the terminal vertex of a_1 coincides with the initial vertex of a_2 - the sink of the directed subgraph induced in A_2 by $\sigma(a_1)$ coincides with the source of the directed subgraph induced in A_2 by $\sigma(a_2)$.

Table 2.3

σ(develop_A)	= {design_A, implement_A, transfer_Peter}
σ(develop_B)	= {design_B, implement_B}
σ(develop_C)	= {design_C, implement_C, dummy}
σ(develop_D)	= {design_D, implement_D}
σ(develop_E)	= {design_E, implement_E}
σ(assembly_AB)	= {assembly_AB}
σ(assembly_(AB)C)	= {assembly_(AB)C}
σ(assembly_DE)	= {assembly_DE}
σ(assembly_(ABC)(DE))	= {assembly_(ABC)(DE)}

Let, for instance, $N_1 = (E_1, A_1, d_1)$ be the activity network of Fig. 2.14, and $N_2 = (E_2, A_2, d_2)$ that of Fig. 2.15. A_1 coincides with the set of activities in Table 2.1, and A_2 with the set of activities in Table 2.2 plus a dummy activity required by Specification 2.2. A_2 is then the subdivision of A_1 corresponding to the split function σ of Table 2.3.

But N_1 is not an unfolding of N_2, because the directed subgraph induced in N_2 by the set of activities σ(develop_A) = { design_A, implement_A, transfer_Peter } has two sinks. And so does the directed subgraph induced in N_2 by the set of activities σ(develop_C).

Even if we cut both activities "transfer_Peter" and "dummy" out of N_2, N_1 does not become an unfolding of N_2.

In that case condition (iii) would be violated by the sequential activities "develop_A" and "develop_B" of N_1. In fact, the sink of the directed subgraph induced by set σ(develop_A) = { design_A, implement_A } in A_2 would then be vertex 6, which does not coincide with vertex 0, source of the directed subgraph induced in A_2 by the set σ(develop B) = { design_B, implement_B }.

3. Disciplined Planning

The human capability of comprehending plans — no matter in which representation language — depends on their size and complexity. The number of elements to be considered and the intricacy of the plan structure may place insurmountable obstacles to the understanding of plans — with or without computer aid. Taking into account all details of very large and complex plans lies outside the range of people's capability. Attempts have been made to apply Artificial Intelligence (AI) techniques to managing large projects, but there are two main difficulties.

First, AI planning systems are based on an exhaustive description of the "world" in which plans are to be generated. In other words, AI planning systems require complete specification of the project development environment. Typically, this specification must include a complete list of allowed "actions", expressed in some IF-THEN equivalent form. It is easy to see that a *complete* specification of the project development environment can well be worked out for toy problems but is unrealizable if we are planning manufacturing or research projects.

Secondly, computer generated plans are output in the form of long, possibly redundant sequences of actions supposed to achieve an input list of goals. Cutting out redundant actions is not the main problem — various strategies have been suggested for that. The true difficulty is the identification, within such a sequence of actions, of functionally relevant macros.

The size and complexity of plans are not reducible whenever they depend on intrinsic features of the project, or on the utilization for which the plan is intended. In these cases, our capability for dealing with large and complex plans can be notably increased by a suitable regulation of the plan development process.

In this chapter we shall introduce an effective approach to the development of plans. It is called *disciplined planning,* and its main guidelines are *top-down development* and *disciplined planning syntax.*

3.1 Top-Down Development of Plans

Though the word project may designate any piece of planned work, project engineering is aimed at projects requiring professional planning, projects which call for the development of large and complex plans.

As a rule, the design of such projects needs the integration of a number of specialized execution plans, concerned with different aspects of project realization: time scheduling, cost evaluation, resource allocation, financial flow control, etc.

To specify these plans at the level of detail ultimately needed for project execution can be very burdensome, but the labor can be brought under control by adopting top-down planning. In this section we will introduce this approach. It is both easy and elegant to do so in the context of activity networks.

Describing is making an abstraction choice. Project specifications are descriptions, and imply a choice of elements to be considered primitive within the framework of specification. This choice precedes the specification itself. It is a design step requiring particular care as it implicitly sets the abstraction level of specification, and consequently that of related plans.

Top-down development of project plans proceeds as follows:

1. The project is subdivided into *a set* S *of subprojects* suitable as primitives in a first "highest level" project specification. Such subprojects will generally require further specification. If this is the case we will call them *macro-activities,* else *activities.*
   Macro-activities must be made coarse enough to make the plan structure intelligible, yet fine enough so that from the managerial level considered, more details are unimportant.

2. On the basis of set S, a *project specification* is worked out.

3. According to the project specification — and hence again on the basis of S — a *plan* will be designed. Often, the project specification will force us to design a set of plans. Suppose, for a start, that one plan has been devised.

4. The plan will be represented in some representation language. We suppose for now that it will be as an activity network $N = (E, A, d)$ where elements of the set A — that is, the activities of N — represent elements of set S — that is, the specified sub-projects. Both labels *activity* and *macro-activity;* — short: *macro* — carry over from elements of S to the corresponding element of A.

5. If all elements of A are activities — that is, if there are no macros — we are done: the activity network N represents a plan suited for the realization of our project. Else we proceed as follows.

6. The activity network N is *unfolded* :

    each macro-activity **m** of N is regarded as an autonomous project, and steps 1 to 5 are repeated on it. This will give for each macro **m** ∈ A an activity network N_m , which we will call a *refinement of* **m**. Substituting in N every arc representing a macro by the corresponding refinement, we get a new activity network N'. It is easy to see that N' is an unfolding of N.

    We rename N' with name N, and go back to step 5.

Top-down development of plans consists in the step-wise generation of a hierarchy of plans. Plans belonging to the same hierarchical level are descendent from plans belonging to the predecessor level by unfolding primitives. Such unfoldings are based on the specification of macros. Therefore, the hierarchy of plans mirrors a hierarchy of specifications. Specification levels correspond to levels of project control. The sharpness of the lowest plan in the hierarchy reflects the sharpness of the lowest level of project management. At each managerial level, qualified plan users must be able to exploit the plan without any further specification.

We will now demonstrate some top-down development steps, carried out on a part of the small project of Specification 2.1.

Suppose we chose three macros — develop_A, develop_B and develop_C — as starting primitives of a part of what Specification 2.1 describes. Our first level specification may then be the following.

Specification 3.1

1. Three program modules — named A, B and C — must be developed.
2. Module A must be developed before module B.

The activity network in Fig. 3.1 represents a plan satisfying Specification 3.1. We declared develop_A, develop_B, develop_C to be macros. This means that for some or all of the plan users further specification will be needed. This would not be the case if, for example, we think develop_A, develop_B, develop_C were tasks of three autonomous software development teams, and plan 3.1 was simply meant for illustrating precedence constraints regarding the work of those teams.

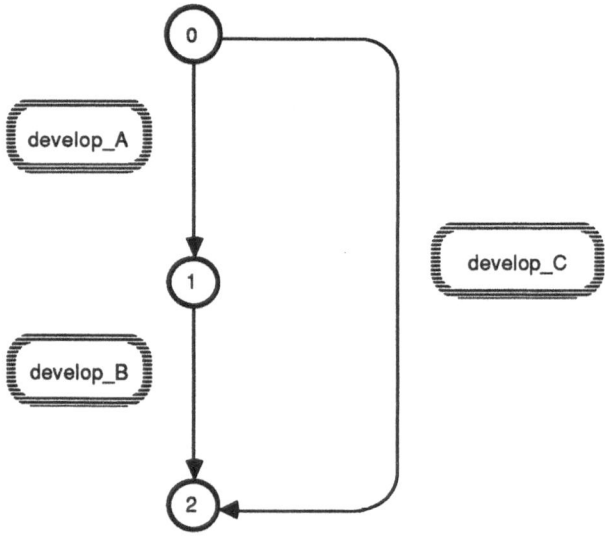

Fig. 3.1

Table 3.1

Macro	Primitives
develop_A	design_M1, implement_M1, design_M2, implement_M2;
develop_B	design_test, implement_test, execute_test;
develop_C	design_M3, implement_M3, design_M4, implement_M4, design_M5, implement_M5.

But develop_A, develop_B and develop_C are macros. And we go on with top-down development. The next step is to specify each of the three macros. We fix the primitives of the second level project specifications as in Table 3.1.

On this basis, the specification of develop_A, develop_B and develop_C could be:

Specification 3.2:

1. develop_A: two program modules — M1 and M2 — must be developed and implemented; the design of a module precedes its implementation;
2. develop_B: a test — to be carried out on the implementation of A — will be first designed, then implemented, and last carried out;
3. develop_C: three program modules — M3, M4, and M5 — must be developed and implemented; the design of a module precedes its implementation.

The activity networks in Fig. 3.2a and 3.2b are refinements of the macros develop_A, develop_B, develop_C. They are representation of plans satisfying points 1 to 3 respectively of specification 3.2.

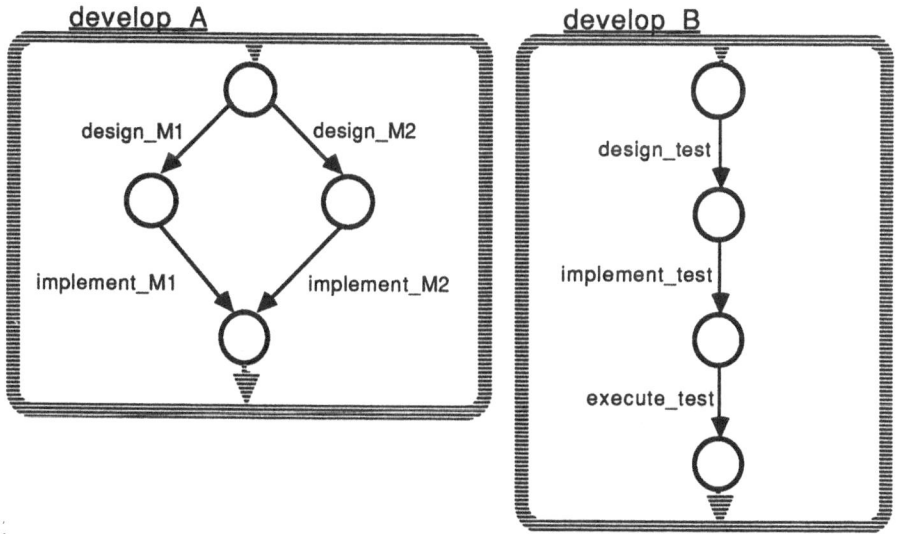

Fig. 3.2a

Substituting a macro by its refinement is carried out by first cutting the macro arc open in any intermediate point. Then the refinement is inserted at that point, and joined to the two macro arc segments. These will inherit the original arc orientation.

We see that the unfolding procedure requires that refinements have exactly one source and one sink. Here we have secured this by asking that refinements be activity networks. The activity network in Fig. 3.3 is an unfolding of that in Fig. 3.1, obtained by means of the refinements illustrated here above. Activity network 3.3 represents a second level plan for our project, a plan devised according to second level specification 3.2, which was in turn formulated on the basis of the choice of subprojects fixed in Table 3.1.

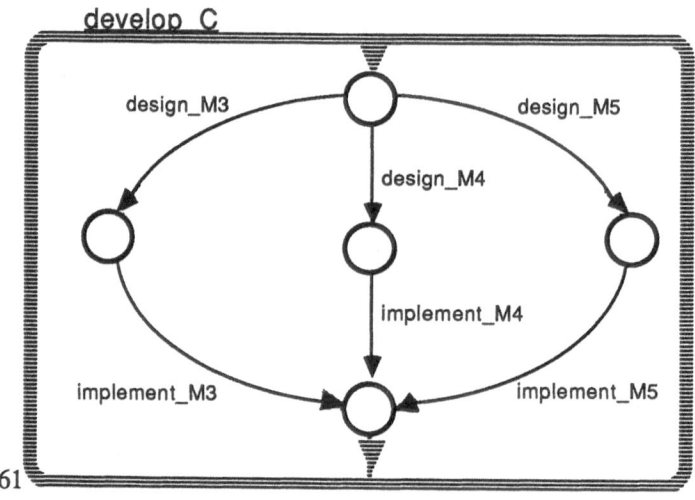

Fig. 3.2b

If plan users do not need any further specification in order to play their role, we are done. Else, we must go on, and fix the primitives of third level project specifications.

After a certain number of unfoldings all specification primitives will be activities. Specification will then lead to the final plan, and top-down development will terminate.

As we mentioned, the final plan is tailored for a specific utilization. Intermediate plans are often usable for other purposes, helpful to different users. The full hierarchy of plans may also have its own interest: it may well be used to structure a hierarchy of responsibilities in the project realization phase.

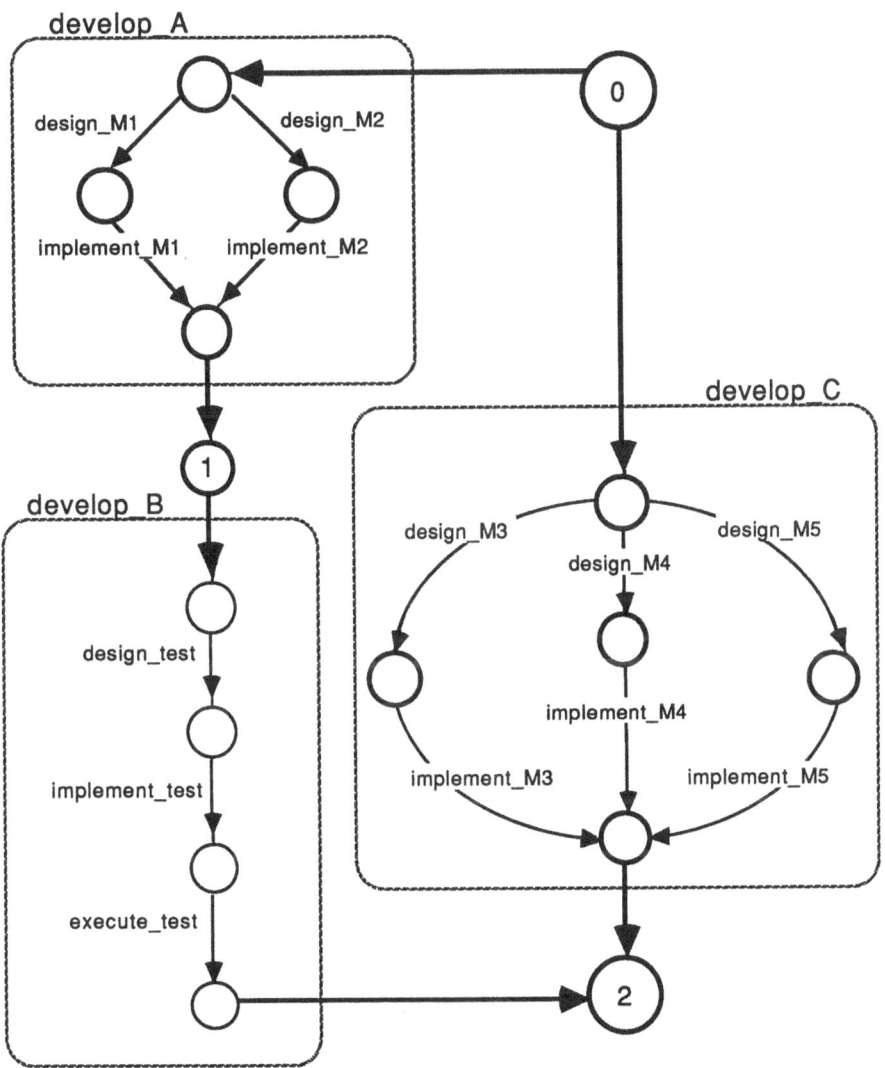

Fig. 3.3

3.2 Disciplined Planning Syntax

The second principle of disciplined planning prescribes a syntax for representing plans on the basis of a set of *composition schemata*. Whatever plan has been conceived, disciplined planning asks that its representation be 'legal' according to the disciplined planning syntax (DPS).

DPS is based on four composition schemata: *sequence, concurrent composition, choice and iteration*. Together, these schemata are sufficient for representing every plan. Plan representations obeying DPS have exactly one entry point and one exit point.

Not all of the four composition schemata apply to all representation languages, but they all — when correctly applied — generate plan representations suited for being directly used as refinement within top-down development.

Getting a 'legal' representation of a plan requires the use of a proper representation language. For instance, ordinary activity networks do not allow us to represent plans involving choice or iteration. But every language supporting the four standard composition schemata is suited for 'legally' representing any plan.

We shall introduce the four standard composition schemata by means of activity networks with cycles and choice nodes — a variant of activity networks. We will neglect activity durations as they do not influence DPS.

DPS consists of the following directives:

1. At any given planning level there are activity units whose work meaning is entirely conveyed by a label and/or accompanying verbal description. Intended plan users will not need any further information in order to understand — or to carry out — the corresponding piece of work.

   Such activity units are called atomic plans. Atomic plans are represented by atomic modules, that is, by activity networks consisting of one labelled arc and its two adjacent vertices.

2. Atomic modules are DPS-correct modules — often we will simply say: correct modules.

   In addition, there are four rules — or schemata — of module composition: sequence, concurrent composition, iteration and choice. Any two correct modules combined according to any one of these four composition schemata yields a correct module.

   This construction procedure applied repeatedly yields all DPS-correct modules.

We will represent DPS-correct constructs with the help of the following graphic conventions: source nodes of plan representations are vertically shaded, sink nodes horizontally. Thick hatched arrows represent correct modules.

3. For any two correct modules **M1** and **M2** the activity network obtained by identifying the sink of **M1** with the source of **M2** is a correct module. This composition schema is called *sequence*.

Figure 3.4 illustrates the composition schema sequence.

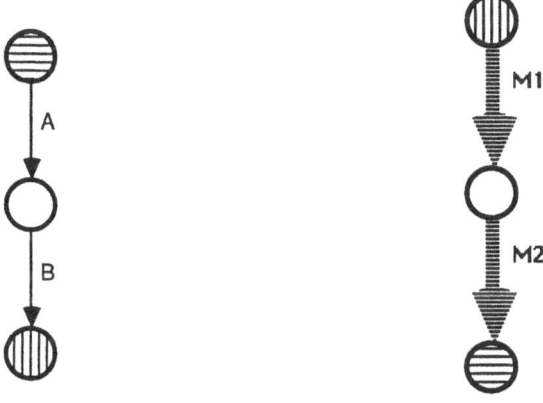

Sequence of Atomic Modules   Sequence of Any Two Correct Modules

Fig. 3.4

4. For any two correct modules **M1** and **M2** the activity network obtained by identifying both the sinks and the sources of **M1** and **M2** is a correct module. This composition schema is called *concurrent composition*.

   For any correct module **M** the activity network obtained by identifying the source and the sink of **M** is a correct module. This composition schema is called *iteration;*.

   Figure 3.5 illustrates the composition schemata concurrent composition and iteration.

5. For any two correct modules **M1** and **M2** the activity network obtained by identifying both sinks of **M1** and **M2** with a choice node is a correct module. This composition schema is called *choice*.

Figure 3.6 illustrates the composition schema choice; the choice node is represented by means of a triangle.

6. Plan representations which are in accordance with the directives 1 to 5 above are said to be correct.

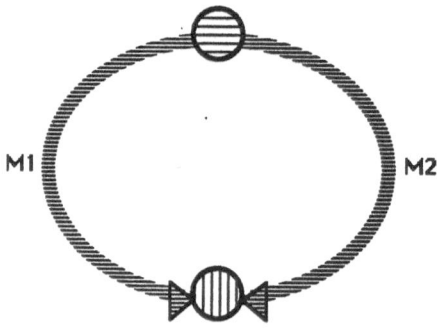

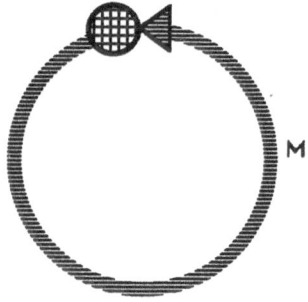

Concurrent Composition of two Correct Modules   Iteration of a Correct Module

Fig. 3.5

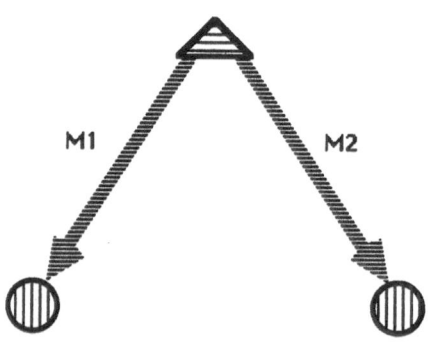

Choice between Two Correct Modules

Fig. 3.6

DPS may be seen as a protocol for the down-top construction of plan representations. Basic correct modules represent atomic plans, and using correct modules further correct modules can be constructed: sequences, concurrent and choice structures, iteration cycles.

The main advantage of DPS is the neat modularity of resulting plan representations — modularity which will allow for the clean distribution of execution responsibilities, and for the easy detection and modification of misbehaving parts of the plan. Plan structure — both logical and causal — becomes transparent, and documentation is better supported.

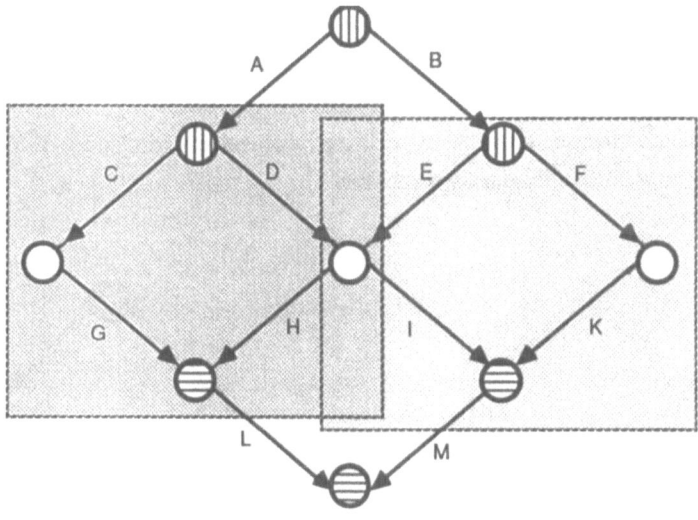

Fig. 3.7

The plan representation in Fig. 3.3 is correct, while that in Fig. 3.7 is not. It would be wrong to regard the plan representation 3.7 as the concurrent composition of the two modules CDGH and EFIK shaded in the figure. Concurrent composition of two modules does not allow one to identify nodes except for the two sources and the two sinks.

Plan representation 3.7 is not correct because the linkage of activities D, H, E and I does not fit into any DPS composition schema. The undesirable consequence of this "lack of discipline" is confusion about causal structure of the plan. Activities belonging to concurrent modules are made sequential: for instance, activities E and H are made sequential although belonging to concurrent modules. The execution independence implied by the concurrent composition of modules CDGH and EFIK is contradicted by the sequentialization of E and H, and of D and I.

Figure 3.8 shows a "disciplined" representation of the same plan, together with a representation where the violation of the plan representation in Fig. 3.7 is preserved but made explicit by the use of dummy activities. The price paid for the "disciplined" representation is giving up precedence between E and H, and between D and I. The gain is the clean separation of causally independent plan parts. This separation will support responsibility subdivision in the phase of execution organization. The execution of one or more concurrent modules can be given to a person in charge without any overlapping of responsibilities.

If precedence between E and H and between D and I cannot be renounced, disciplined development asks us to express the violation above explicitly, using dummy activities. This puts a clear label on the dependence/independence ambiguity of the plan, and hence on control difficulties during execution.

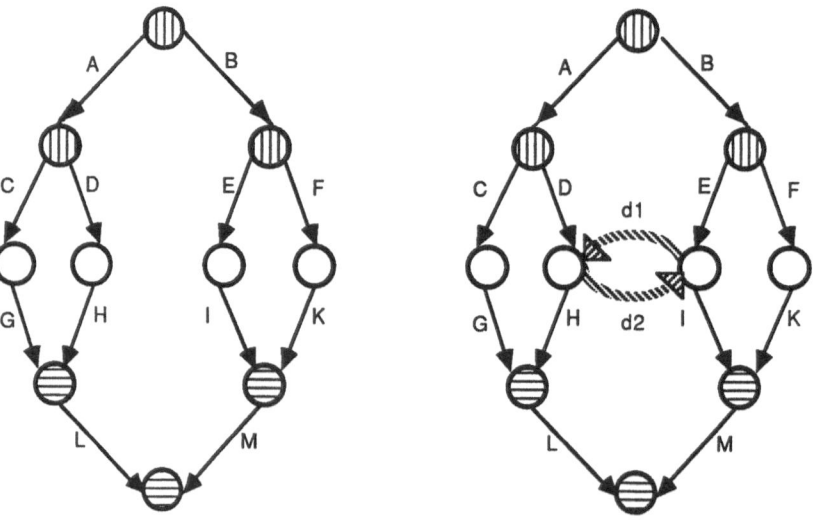

Fig. 3.8

Another possible way to get a disciplined plan representation is the direct modification of project specification. Representation of two different plans — both deriving from
modified project specifications — are shown in Fig. 3.9.

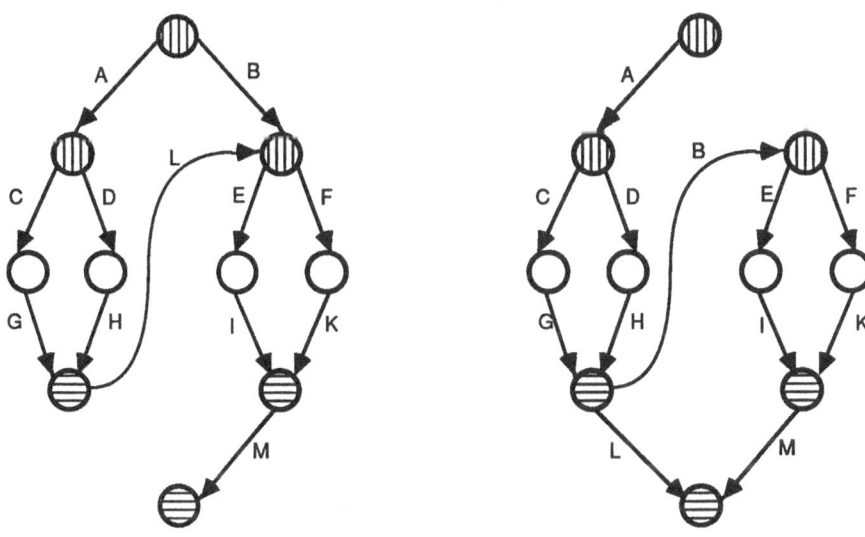

Fig. 3.9

The use of dummy activities — and in general "wild" planning — recalls the use of
the GOTO statement in programming, both in the way it works — lawless jumps out of
standard modules — and in the effect it produces — loss of logical clarity. As with GO-
TOs, there may sometimes be good reasons for renouncing disciplined planning, but this
should in any case be done consciously.

The next chapter is dedicated to the formal introduction of the popular and widely
used project representation and analysis techniques CPM, PERT and GERT.

4. Planning under Precedence/Duration Constraints: Networking Techniques

The networking techniques we will present in this chapter — CPM, PERT and GERT — allow us to manage essential planning problems: scheduling, cost planning and resource allocation. Also, sticking to disciplined planning is natural and easy, since the representation languages used are variants of activity networks.

CPM (Critical Path Method) was developed at the end of the fifties by Walker of the DuPont Company and Kelley of the Univac Division of Remington Rand Corporation as a tool for programming maintenance activities in DuPont industrial plants.

PERT (Program Evaluation and Review Technique) was developed independently during the same years at the U. S. Navy's Special Projects Office. It grew from planning and scheduling the realization of the Polaris missile. CPM and PERT are rather similar techniques, so that they are often referred to with the common name of CPM-PERT. But we will see that there are good reasons for introducing them separately.

GERT (Graphical Evaluation and Review Technique) was first introduced by Pritsker, in the middle sixties, for planning and analyzing terminal count-down of an Apollo space system.

4.1 No Choices, No Cycles, Known Durations: CPM

The Critical Path Method offers a technique for the design and control of projects requiring development of large and complex plans.

Activities take place over time, and require resources as they progress. Large projects will hardly meet their objectives without careful planning of times, costs and resources. The CPM method helps whenever project plans can be assumed to be free of choices and cycles, and when the planner is able to estimate the duration of activities.

CPM has found wide acceptance in business and administration environments for the simplicity of its graphical representation language and the direct applicability of its outcome.

The Critical Path Method only applies to plans that are free of choices and cycles because it is based on the activity network representation of plans.

The project is subdivided into a finite set S of non-overlapping, partially ordered *subprojects* to be considered primitives within a first project specification. These subprojects will often require further specification. If so, we will call them *macro-activities,* else *activities*.

On the basis of S, a *project specification,* and hence a *project plan* — or a set of project plans — is worked out. This plan will be expressed in terms of *macro-activities* and *activities* , and will mirror the partial ordering of S. In this chapter, we shall only consider plans with no macro-activities.

CPM requires that the considered project plan is represented as an activity network $N = (E, A, d)$ such that the elements of A represent the elements of S — that is, the initial subprojects. We shall apply the denomination *activity* both to the elements of S and of A. The partial ordering among the planned activities will give the precedence relation of the network.

Not every partial ordering of activities can be directly expressed by means of an activity network. Sometimes, dummy activities will be necessary in order to represent the plan activity ordering fully.

CPM requires that duration of planned activities to be estimated by means of a real number. This number will give the value of function d for the activity.

Execution Control: Event Time Bounds

In activity networks, activities are given a constant duration, and the output activities of events are assumed to start as soon as the event is realized. This behavior describes the "ideal" — the planned — project execution.

No actual plan execution will behave like that. Duration of activities may be shorter or longer than planned. At event realizations, the output activities will start neither immediately nor all together. But straying from the plan is not always an undesired happening. Controlling the start or duration of activities is also a way for improving the allocation of resources.

CPM supports the supervision of plan executions. Start and duration of activities can be controlled within the bounds set by two characteristic event constants: the earliest event time and the latest event time.

We assume that project execution starts at time zero. The earliest time point at which the project can be completed is called the *project completion date* .

The earliest time an event $e = (\text{In, Out})$ can take place is the time point at which all activities of In are first completed. This date is called the *earliest event time* , or the *realization date* of e — we denote it by $E(e)$.

The latest time an event e can take place without delaying the project completion date is called the *latest event time* of e — we denote it by $L(e)$.

We will call the difference between the latest and the earliest event times of event e the *slack* of e . We will denote it by $S(e)$.

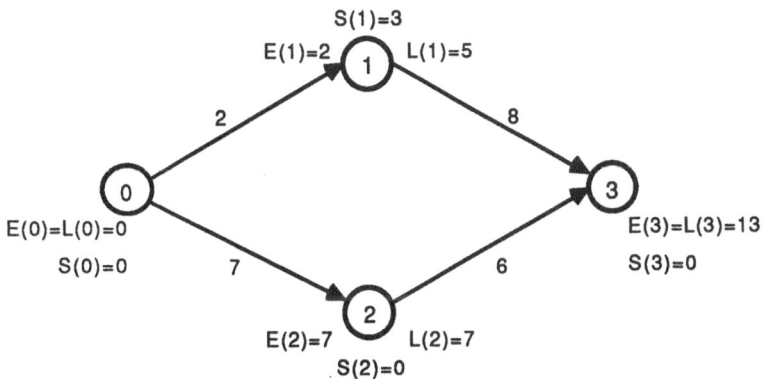

Fig. 4.1

Consider, for instance, the activity network in Fig. 4.1.

Arcs represent activities. Integers at the arcs express duration of the corresponding activities.

The earliest and latest event times of the source event are both 0, by convention. The earliest and latest event times of the sink event also coincide, by definition.

We see by inspection that the earliest termination of the input activities of event 3 is 13. Therefore $E(3) = 13$.

The execution manager may protract the duration of activity $0 \rightarrow 1$ until time point 5 without affecting the project completion date. Or he could defer the start of activity $1 \rightarrow 3$ up to time point 5.

Also, he could think out suitable combinations of duration and start slackening. This would allow him to control the execution of the planned activities, and stick to the agreed time schedule.

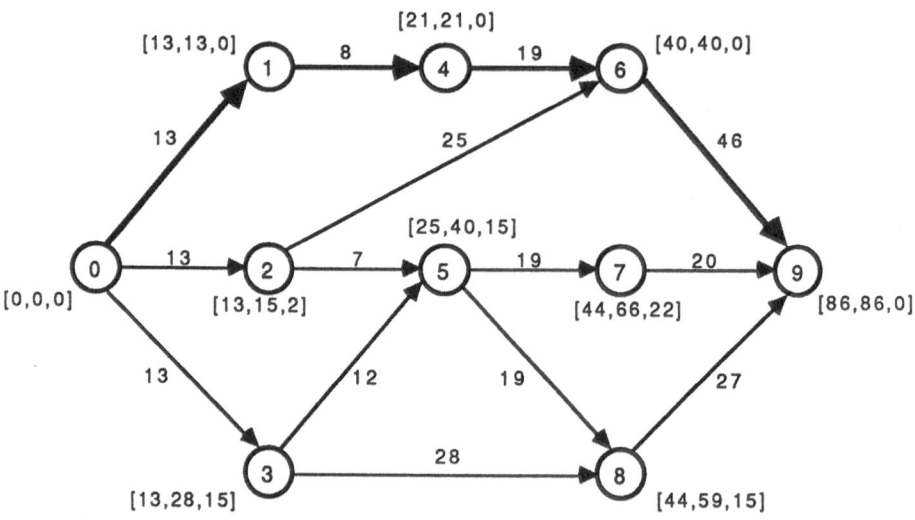

Fig. 4.2

Formally, let $N = (E, A, d)$ be the activity network representing a certain plan. Then the *earliest (event) time* of event e is defined by:

(i) $E(e) := 0$ if e is the source
(ii) $E(e) := \max_{i \to e}[E(i) + d(i \to e)]$ else
        where $d(i \to e)$ denotes the duration of activity $i \to e$;

the *latest (event) time* of event e is defined by:
(i) $L(e) := \max_{i \to e} E(i)$ if e is the sink
(ii) $L(e) := \min_{e \to j}[L(j) - d(e \to j)]$ else
        where $d(e \to j)$ denotes the duration of activity $e \to j$;

the *(event) slack* of e is:
        $S(e) := L(e) - E(e)$.

In the CPM plan of Fig. 4.2 triplets close to vertices list the earliest time, the latest time and the slack of the corresponding event — in this order.

We see that activities $5 \to 7$ and $5 \to 8$ can be started between date 25 and date 40. Activity $5 \to 7$ must be completed between date 44 and date 66, activity $5 \to 8$ between date 44 and date 59. An execution supervisor can shift or prolong the execution of these activities within the time intervals above. Activity $5 \to 7$ has a little more execution flexibility than activity $5 \to 8$.

Activity $4 \to 6$ has no execution flexibility: it must start at date 21, and end at date 40. Neither shifting nor prolongation are allowed.

Critical Paths

In any activity network $N = (E, A, d)$, for all events e it holds true that $L(e) \geq E(e)$.

Events for which $L(e) = E(e)$ — or, equivalently, $S(e) = 0$ — are called *critical events*.

Each path from source to sink exclusively made up of critical events is called a *critical path*.

Activity networks have at least one critical path. Figure 4.3 shows an activity network with two critical paths: $0 \to 1 \to 4 \to 6 \to 9$ and $0 \to 3 \to 8 \to 9$. Activities lying on a critical path are called *critical activities*. It is very easy to determine critical paths by tracing from source to sink exclusively over zero slack events.

Critical paths have *maximum duration* over all paths from source to sink: there is no time to spare for the execution of critical activities. This maximum duration is the minimum time interval in which the plan can be executed. Any delay in the execution of a critical activity will put off the project completion date. Critical activities must therefore be most tightly controlled.

In the CPM network of Fig. 4.2: critical events are **0, 1, 4, 6** and **9**; the only critical path is **0→1→4→6→9**; the minimum time interval in which the project can be completed is 86 time units.

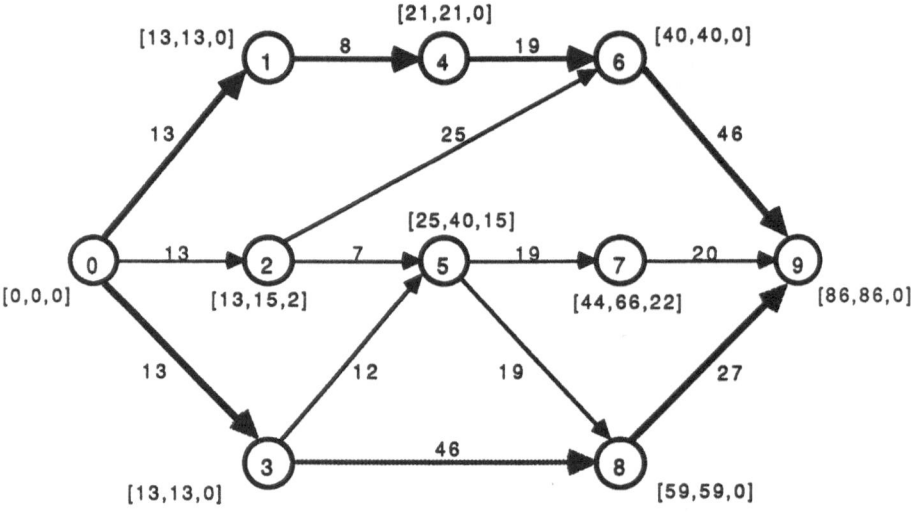

Fig. 4.3

Event Realization Slacks

Non-critical events may be delayed without affecting the project completion date. The flexibility of the time schedule can be evaluated considering the planned delays of non-critical events.

Three characteristic constants must be evaluated: the slack S — already introduced — the free slack FS and the independent slack IS.

The *slack* of an event **e** is the longest time interval the realization of e can be postponed from its earliest time without delaying the project completion date.

The *free slack* of an event e is the longest time interval its realization can be postponed from earliest time without preventing any next event from being realized at its earliest time.

The *independent slack* of an event e is the longest time interval the realization of e can be postponed from its earliest time independently of the realization date of immediately preceding and subsequent events.

In other words, the independent slack is the longest delay of e under the worst scheduling conditions — that is, when all immediately preceding events are realized at the latest time, and all directly subsequent events are realized at the earliest time.

Event slack S, free slack FS and independent slack IS are equal to zero for critical events.

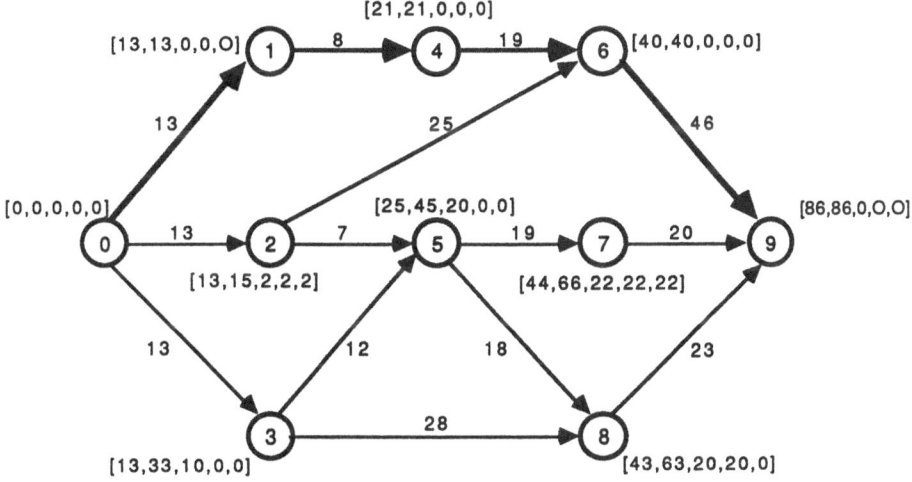

Fig. 4.4

Formally: let N = (E, A, d) be the activity network representing a certain plan, and e a non-critical event of N. The *free slack* of e is defined by
$$FS(e) := \min_{e \to j} [E(j) - d(e \to j)] - E(e),$$
the *independent slack* of e is defined by

$$IS(e) := \max \{ 0, \min_{e \to j} [E(j) - d(e \to j)] - \max_{i \to e} [L(i) + d(i \to e)] \}.$$

Recall that we have already defined the *slack* of a generic event e as

$$S(e) = L(e) - E(e).$$

We will now illustrate the computation of the delay constants for some non-critical events of the plan represented in Fig. 4.4. The vectors near the vertices list the values of E, L, S, FS, IS, for the corresponding event.

Table 4.1

event	E	L	S	FS	IS
0	0	0	0	0	0
1	0	0	0	0	0
2	13	15	2	2	2
3	13	33	10	0	0
4	0	0	0	0	0
5	25	45	20	0	0
6	0	0	0	0	0
7	44	66	22	22	2
8	43	63	20	20	0
9	0	0	0	0	0

$$S(2) = 2$$
$$FS(2) = \min_{j \in \{5,6\}} [E(j) - d(2 \to j)] - E(2) = \min \{40 - 25, 25 - 7\} - 13 = 2$$
$$IS(2) = \max \{0, \min_{j \in \{5, 6\}} [E(j) - d(2 \to j)] - [L(0) + d(0 \to 2)]\}$$
$$= \max \{0, \min\{40 - 15, 25 - 7\} - [0 + 13]\} = 2$$

$$S(5) = 20$$
$$FS(5) = \min_{j \in \{7,8\}} [E(j) - d(5 \to j)] - E(5) = \min\{25, 25\} - 25 = 0$$
$$IS(5) = \max \{0, \min_{j \in \{7, 8\}} [E(j) - d(5 \to j)] - \max_{i \in \{2, 3\}} [L(i) + d(i \to 5)]\}$$
$$= \max \{0, \min\{25, 25\} - \max\{15 + 7, 33 + 12\}\} = 0$$

$$S(8) = 20$$

FS(8) = [E(9) − d(7→9)] − E(8) = 66 − 44 = 22

IS(8) = max { 0, [E(9) − d(7→9)] − [L(5) + d(5→7)] } = 2

S(9) = 15

FS(9) = [E(9) − d(8→9)] − E(9) = 63 − 43 = 20

IS(9) = max { 0, [E(9) − d(8→9)] − max$_{i \in \{3, 5\}}$ [L(i) + d(i→8)] } = 0

Table 4.1 summarizes the CPM delay constants for the project plan represented in Fig. 4.4. Critical events are underlined.

Activity Time Bounds

The parameters introduced above all relate to events. But their information content can also be expressed in terms of activities. This mode is very popular in applications.

The reason is that we mostly speak about plans in terms of activities: people are usually concerned with the beginning, termination or delay of activities, and not with the occurrence or delay of events.

The activity related constants of the CPM time scheduling technique are the time points of earliest and of latest activity start, and the time point of earliest and of latest activity completion. We introduce them now.

Let N = (E, A, d) be the activity network representation of a certain plan, and let i→j be any activity of N.

The *earliest start* of activity i→j is defined by

ES(i→j) := 0   if i is the source

ES(i→j) := max$_{h \to i}$ [ES(h→i) + d(h→i)]   else.

The *earliest termination* of activity i→j is defined by:

ET(i→j) := d(i→j)   if i is the source

ET(i→j) := max$_{h \to i}$ ET(h→i) + d(i→j)   else.

It is easy to see that ES(i→j) = max$_{h \to i}$ ET(h→i).

The *latest termination* of activity i→j is defined by:

LT(i→j) := max$_{h \to j}$ ET(h→j)   if j is the sink

LT(i→j) := min$_{j \to k}$ [LT(j→k) − d(j→k)]   else.

The *latest start* of activity i→j is defined by:

$$LS(i{\rightarrow}j) := \max_{h{\rightarrow}j} ET(h{\rightarrow}j) - d(i{\rightarrow}j) \qquad \text{if } j \text{ is the sink}$$
$$LS(i{\rightarrow}j) := \min_{j{\rightarrow}k} LS(j{\rightarrow}k) - d(i{\rightarrow}j) \qquad \text{else.}$$

Notice that if j is the sink $LT(i{\rightarrow}j)$ is by definition the earliest project completion date. Obviously:

$$LS(i{\rightarrow}j) = LT(i{\rightarrow}j) - d(i{\rightarrow}j).$$

Characteristic delay constants are defined for activities too: total delay, free delay, linked delay and independent delay.

Table 4.2

activity	duration	ES	ET	LS	LT	D	FD	ID
0→1	13	0	13	0	13	0	0	-
0→2	13	0	13	2	15	2	0	-
0→3	13	0	13	20	33	20	0	-
1→4	8	13	21	13	21	0	0	0
2→5	7	13	20	38	45	25	5	3
2→6	25	13	38	15	40	2	2	0
3→5	12	13	25	33	45	20	0	0
3→8	28	13	41	35	63	22	2	0
4→6	19	21	40	21	40	0	0	0
5→7	19	25	44	47	66	22	0	0
5→8	18	25	43	45	63	20	0	0
6→9	46	40	86	40	86	0	-	-
7→9	20	44	64	66	86	22	-	-
8→9	23	43	66	63	86	20	-	-

The *total delay* of an activity is the longest time interval the activity termination can be postponed from the earliest termination date without delaying the project completion date.

The *free delay* of an activity is the longest time interval the activity termination can be postponed from the earliest termination date without preventing all directly subsequent activities from taking place at respective earliest start dates.

The *independent delay* of an activity is the longest time interval its actual start (termination) date can be postponed from its earliest start (termination) date independently of the realization dates of immediately preceding and immediately subsequent activities. In other words, the independent delay is the longest delay, under the assumption of the most unfavorable scheduling conditions: each immediately preceding activity completed at its latest termination date, and each directly subsequent activity begun at its earliest starting date.

All delay constants above are equal to zero for critical activities.

Formally — and with the same notations as above:
the *total delay* of an activity $i \rightarrow j$ is
$$D(i \rightarrow j) := LT(i \rightarrow j) - ET(i \rightarrow j),$$
the *free delay* of an activity $i \rightarrow j$ with j other than the sink is
$$FD(i \rightarrow j) := \min_{j \rightarrow k} ES(j \rightarrow k) - ET(i \rightarrow j),$$
the *independent delay* of an activity $i \rightarrow j$ with j other than the sink is
$$ID(i \rightarrow j) := \max \{ 0, \ \min_{j \rightarrow k} ES(j \rightarrow k) - \max_{h \rightarrow i} LT(h \rightarrow i) - d(i \rightarrow j) \}.$$

Table 4.2 displays the delay analysis of Table 4.1 in terms of activities. Critical activities are underlined.

4.2 No Choices, No Cycles, Known Direct Costs: Cost-CPM

In this section we will show how CPM supports cost planning. The same method also applies to PERT networks. A strong point in favour of both CPM and PERT is the systematic base they provide for the consistent development of time schedules and cost plans.

Cost planning starts with a set of initial, rough feasibility estimates, and evolves to more and more accurate cost plans set up during subsequent project development stages. In the project execution phase the costs of not yet completed activities will in general require to be re-planned.

In the sequel we will be concerned with the so-called direct costs, that is, with costs which can be accounted to activities on a per-activity basis. Therefore, the cost centers will be the activities.

Which costs are to be considered direct depends on the managerial level of the cost plan. If the whole project is represented as a single activity, every cost must be accounted to this activity — every cost will be direct. As the plan becomes more detailed, some costs will be overhead to groups of activities — or to all of them — and will no longer appear in the cost plan.

CPM-PERT cost planning is based on the interplay of three financial aggregates: planned project cost, nominal cost, actual cost.

The *planned project cost* is the sum of all activity costs initially forecasted on the basis of an agreed-on project time schedule. Such costs are called *planned costs*.

The *planned cost at date* ∂ is the amount of forecasted activity costs at time point ∂.

The *nominal cost at date* ∂ is the amount which would have been paid at time point ∂ if all costs were paid at the moment of actual termination of the activity they relate to, and were exactly those forecasted in the cost plan.

The *actual cost at date* ∂ is the amount paid at time point ∂ if all costs were paid at the moment of actual termination of the activity they relate to.

Consider the plan represented by Fig. 4.4 and its delay analysis in Table 4.2. Let the planned cost of each activity be that given in Table 4.3. (Duration and cost entries refer to opportune units.)

If we agree to start activities at their earliest start and to consider costs to be matured at earliest termination of the corresponding activity, then the planned project cost is graphically represented by the diagram in Fig. 4.5.

If the time unit is the week and the cost unit the dollar, this diagram forecasts project completion in 86 weeks with a cost of $36000.

Now, suppose the project execution is controlled at the end of week 30. Activities with earliest termination before or at the end of week 30 should have completed, while other activities should be ongoing. According to our time schedule, activities $0{\rightarrow}1$, $0{\rightarrow}2$, $0{\rightarrow}3$, $1{\rightarrow}4$, $2{\rightarrow}5$ and $3{\rightarrow}5$ should have terminated, while activities $2{\rightarrow}6$, $3{\rightarrow}8$, $4{\rightarrow}6$, $5{\rightarrow}7$ and $5{\rightarrow}8$ should be ongoing.

The planned cost at the end of week 30 is $19600.

Table 4.3

activity	duration	prec. activities	planned cost
0→1	13	-	2500
0→2	13	-	2000
0→3	13	-	2800
1→4	8	0→1	8000
2→5	7	0→2	1500
2→6	25	0→2	1200
3→5	12	0→3	2800
3→8	28	0→3	3000
4→6	19	1→4	1400
5→7	19	2→5, 3→5	3500
5→8	18	2→5, 3→5	2800
6→2	46	2→6, 4→6	1200
7→9	20	5→7	1800
8→9	23	3→8, 5→8	1500

Project execution control will in general expose some delay: activities which should have terminated will have failed to do so. If this is the case, the nominal cost at that date will be smaller than the planned cost.

Indeed, if for instance activities 2→5 and 3→5 did not terminate at the end of week 30, the nominal cost at that date will be $15300, that is, $19600 minus the planned cost of activities 2→5 and 3→5.

A positive difference between planned and nominal costs at a certain date indicates that at that time point less work has been accomplished than scheduled.

Also the actual cost at date ∂ will in general be different from nominal cost. This difference indicates that matured costs turned out to be different from planned costs: resources have been more costly, or less productive, or both.

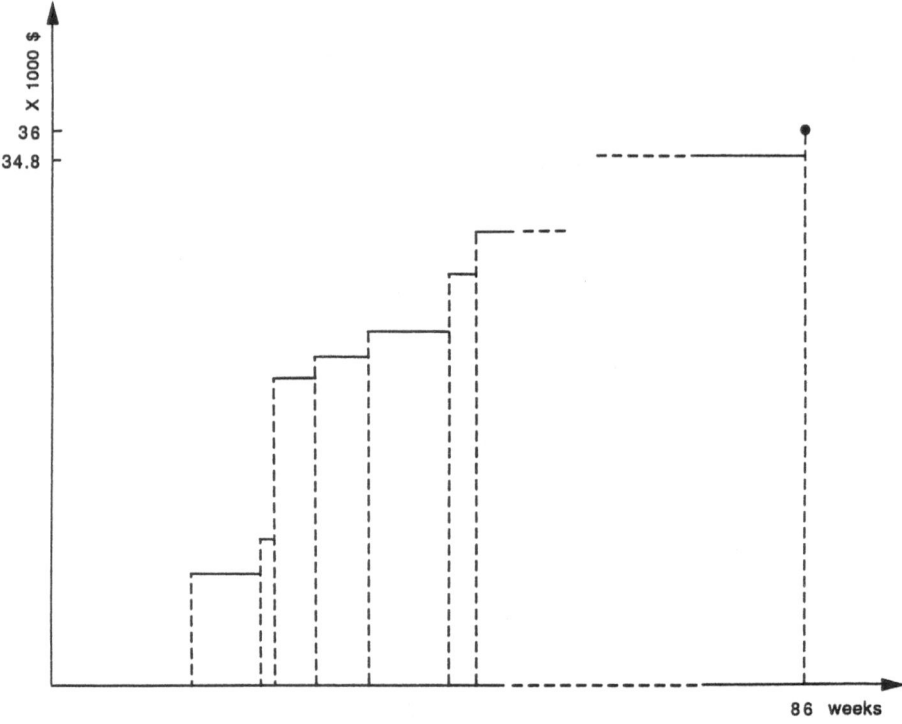

Fig. 4.5

Updating of Time Schedule and Cost Plan

Whenever the project execution control exposes some delay on the critical path, the remaining part of the plan needs to be *re-scheduled*. Even though there might be no reason
for inserting or deleting activities, a new CPM network will be set up. A new source node
is introduced in order to represent the — unique — start point of the remaining plan. This
new source is obtained by identifying the initial vertices of arcs representing not yet com-

pleted activities. The remaining duration of such activities must be estimated and associated with the corresponding arcs. The precedence relation may require re-definition.

A new project completion date will be computed, and the cost plan updated on the basis of the new time schedule.

A common way to update the planned costs of remaining activities is to multiply each of them by the so-called *updating coefficient*. This is the ratio of actual cost to nominal cost, calculated at the date of project execution control.

Table 4.4

activity	duration	prec. activities	updated cost
2→5	2	-	1764
2→6	8	-	1411
3→5	4	-	3293
3→8	11	-	3528
4→6	10	-	1646
5→7	19	2→5, 3→5	4116
5→8	18	2→5, 3→5	3293
6→9	46	2→6, 4→6	1411
7→9	20	5→7	2117
8→9	23	3→8, 5→8	1764

Let us now demonstrate the updating of time schedule and cost plan by means of the activity network plan of Fig. 4.2.

The updating coefficient at the end of week 30 is 1.176, given that $18000 is the actual cost at the same date. Recall that nominal cost at that time point is $15300. Table 4.4 lists updated values of planned costs.

The remaining duration of activities which should already have terminated at the end of week 30 but have not is estimated by the person in charge. By plan ongoing activities are assumed to have the forecasted life.The updated CPM diagram is shown in Fig. 4.6, where the network source is obtained folding 'old' vertices 2, 3 and 4 together.

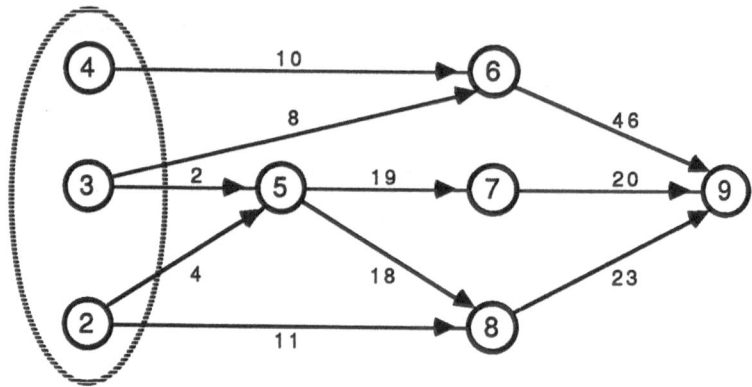

Fig. 4.6

Table 4.6

activity	ES	ET
2→5	0	2
2→6	0	8
3→5	0	4
3→8	0	11
4→6	0	10
5→7	4	23
5→8	4	22
6→9	10	56
7→9	23	43
8→9	22	45

We agree as above that all activities begin at their earliest start date, and that costs mature at the earliest termination date of the corresponding activity. Table 4.6 lists the updated values of the earliest start and termination of the activities to be carried out.

The new forecast is to complete project execution in 86 weeks with a cost of $ 42343. The project completion date did not change since the actual delay of both activities

2→5 and 3→5 was absorbed by the respective total delay. The cost increment depends on the fact that at the end of week 30 the actual project cost is greater than the nominal one.

Cost Optimization

ˉTp to now activity costs were estimated by means of reals. But if cost optimization is within the scope of our project analysis, more information will be necessary.

Activity durations will then be assumed to be non-negative real variables, and the cost of a single activity to be a function of duration — a function shaped as shown in Fig. 4.7.

The planner is supposed to know the cost function of each activity, and in particular the duration value d_0 at which the activity cost is minimum. Activity durations are assumed to have a positive minimum d_{min} corresponding to the maximum cost c_{max}.

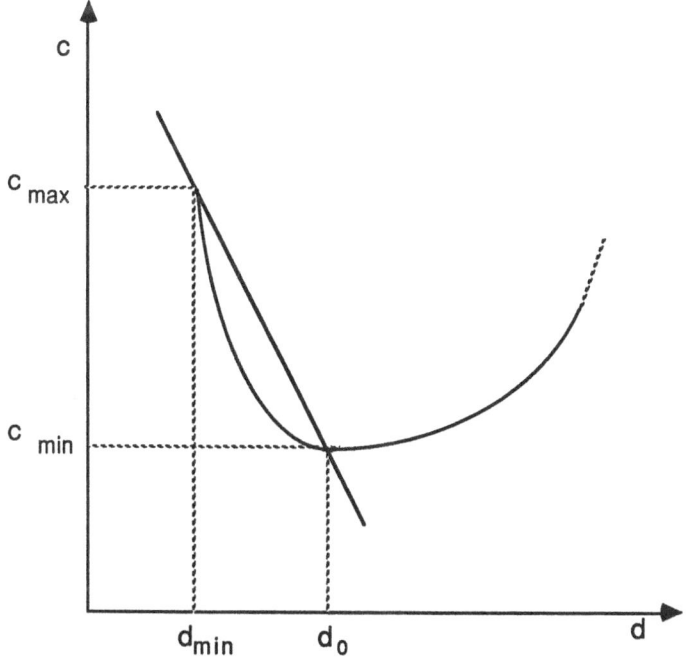

Fig. 4.7

Let function 4.7 be the cost function of a certain activity a, and c_{min} its minimum value, corresponding to d_0.

Obviously, d_0 will be the duration initially associated with activity a.

When an initial duration value is associated with each network activity, the project completion date can be calculated. If this date falls later than required by the project management, the setting of initial activity durations must be modified.

Such an adjustment always implies a positive cost, to be kept minimal. An optimal setting of activity durations — that is, project completion date adjustment with minimal cost — can be found out either analytically or heuristically.

In the next two sections we will introduce both approaches assuming virtually unconstrained resource availability.

Analytical Cost Optimization Algorithm

Analytical cost optimization is based on the application of linear programming methods to the determination of optimal activity duration settings.

At first glance, the algorithm below may look more rigorous and reliable than it is. It is worth noticing that often, in practice, both evaluation of durations d_0 and of corresponding values c_{min} turn out to be matter of opinion. On the other hand, approximation of activity cost by means of a linear function of duration is indeed a strong simplification.

The computational effort is a minor drawback if suitable computation resources are available.

Assume we have a CPM network representation of a certain project plan, and set the initial duration d_0 of each activity as described above. Since, for each activity, the initial duration d_0 corresponds to the minimum cost, the project cost is minimum.

Suppose now the setting of initial activity durations d_0 gives a project completion date ∂ which falls later than the requested project completion date ρ.

In this case, we must work out a different setting of activity durations giving a new planned project completion date $\partial' \leq \rho$. This must be done keeping project cost minimal.

The linear programming formulation of the problem above is the following.

$\Diamond$ Associate to each activity a:

- an interval $[d_{min,a}, d_{0,a}]$ such that $d_{min,a}$ is the minimum duration and $d_{0,a}$ the initial duration of a,
- a linear function $c_a = h_a d_a + k_a$ where the independent variable d_a expresses the unknown duration of a, the dependent variable c_a the planned cost of a, and h_a and k_a are constants chosen so that the straight line $c_a = h_a d_a + k_a$ passes through points $(d_{min,a}, c_{max,a})$ and $(d_{0,a}, c_{min,a})$.
◊ Calculate the values $d_{a_1}, d_{a_2}, ..., d_{a_n}$ which minimize the project cost function

$$C = \Sigma_a c_a = \Sigma_a [h_a d_a + k_a]$$

where a ranges over the set of network activities, under the $n+1$ constraints

$$d_{min,a} \le d_a \le d_{0,a} \quad , \quad \partial' = \rho$$

where ∂' is the project completion date given that $d_{a_1}, d_{a_2}, ..., d_{a_n}$ is the setting of durations, and ρ the requested project completion date.

This formulation is typical for linear programming problems. The problem has a non-trivial solution if and only if $\rho \ge \mu$, where μ is the planned project completion date if the minimum duration is associated with each activity.

If $\rho \ge \mu$, the problem can be solved by means of classical linear programming algorithms — for instance, by means of the simplex method.

A Heuristic Speeding Up Algorithm

Let $N = (E, A, d)$ be an activity network plan for a certain project, ∂ the planned project completion date, and ρ the requested completion date.

If $\rho < \partial$, re-assignment of durations to activities is necessary in order to anticipate the project completion date. This must carried out keeping the speeding up cost under control. In practice, heuristic algorithms are used to this end.

The speeding up algorithm we present here is quite effective in keeping the adjustment cost low, even though it does not always lead to an optimal duration assignment. It is assumed that N has exactly one critical path.

Heuristic Speeding Up Algorithm

- We associate with each activity $a \in A$ its minimum duration $d_{min,a}$;
- we compute the project completion date μ corresponding to this assignment of durations;

{ μ is the minimum project completion date }

- if μ is greater than the requested project completion date ρ, the problem has no solution, and we quit;

{ the project is not realizable within the requested time bounds }

- else: we associate to each activity a its initial duration $d_{0,a}$, and compute the corresponding project completion date ∂;

- if $\partial \leq \rho$, the problem is solved and we go to **;

{ the actual assignment of durations to activities is the required one }

* else:

we set $k := 0$ and $\sigma := \partial - \rho$;

{ σ is the total required project shortening }

we find the set B of critical activities which can be shortened;

{ notice that B is not empty since $\mu \leq \rho < \partial$ }

- for each activity $b \in B$ we compute the *possible shortening*

$$s_b := d_{0,b} - d_{min,b}$$

and the *unitary speeding up cost*

$$u_b := (c_{max,b} - c_{min,b}) / s_b ;$$

- we rename the elements of B as $b_1, b_2, \ldots, b_r$ so that

$$b_i < b_j \text{ if and only if } u_{b_i} \leq u_{b_j} ;$$

* we increment k by 1;

we compute the *actual shortening* α_{b_k} of activity b_k :

$$\alpha_{b_k} := \min \{ \sigma_{b_k}, \sigma - \Sigma_h \alpha_{b_h} \}$$

where $h \in \{1, 2, \ldots, k-1\}$ if $k > 1$, and $h \in \emptyset$ else;

{ the actual shortening of activity b_k is the minimum between its possible shortening and the remaining total shortening $\sigma - \Sigma_h s_{b_h}$ }

- we shorten the duration of activity b_k by α_{b_k}:

$$d_{b_k} := d_{0,b_k} - a_{b_k} ;$$

- we compute the new project completion date ∂';

- if $\partial' > \rho$, we find the set B of critical activities which can still be shortened, and go back to point *;

- otherwise, the problem is solved;

{ the actual assignment of durations to activities is the required one }

** we compute the project cost C, and we end up.

Operational Considerations

The CPM scheduling planning technique produces two sorts of results: a graphical result — in substance, the activity network representation of the project plan and the related critical paths — and numerical results, consisting in the values of event times, event slacks and activity delays.

Making the network representation of a large plan sufficiently readable and expressive requires experience and often a fairly artistic bent, even if accomplished with computer aid.

The graphical representation of the activity network is in principle not necessary in order to get the numerical output above: the computation of times, slacks and delays only requires a suitable listing of activities, durations and follower activities. Therefore, the use of such a listing without graphics makes sense if the complexity of the project will cause the problem of graphical representation to get out of hand.

But there aren't many project engineers willing to renounce all graphical plan representation. A reason for this generalized attitude is the common experience that the effort of constructing the graphical representation has a positive feed-back effect on plan understanding, and improves logical clarity. Given, of course, that representation language and development discipline are adequate. Graphical plan representations always bring causal relationships among plan components to light — even though often restricted to temporal or logical relationships.

The CPM graphical representation language is convenient if the considered plan does not require representation of choices and cycles, and if the planner can represent activity durations by means of a single number. When cost optimization is required, the planner must be able to provide a suitable cost function for each activity.

Once the decision is made to apply the CPM technique for designing and analyzing a certain project, one proceeds according to following guidelines:
- The project will be subdivided into a finite set S of non-overlapping partially ordered *subprojects* , to be considered primitives within the first project specification.
  The *managerial level* at which such first level plans are to be used will determine this subdivision.
- The subprojects will be declared to be either *macro activities* requiring further specification or simply *activities*.
- On the basis of S a *first level project specification* is worked out. In particular, the *duration* of each activity is assigned.

If the plan use includes cost analysis, the direct *cost* of each activity must be fixed. If cost optimization is required, the cost function of each activity has to be provided too.

- A *first level project plan* — more often a set of project plans — will be worked out. These plans will be expressed in terms of *macro activities* and *activities* , and will reflect the partial ordering of S.

- Suppose one plan was devised. This plan will be represented as an *activity network* $N = (E, A, d)$ constructed in accordance with PRS — the plan representation syntax.

- More detailed plans will in general be necessary in order to start project execution — plans for different purposes and uses.
  We will proceed with *top-down plan development* until executable plans are obtained.

- The CPM *time scheduling* technique may be applied to all of these plans.
  With reference to different managerial levels, project completion dates will be calculated; time constraints of intermediate events will be determined together with the critical activities; execution slacks will be computed and analyzed.

- If required, a *cost plan* will be worked out. This may lead to project review, and possibly to project abortion.

- If activities must be speeded up in order to get an acceptable project completion date, the *speeding up cost* must be evaluated. This too, may cause project abortion.

- When a satisfactory cost plan has been set up, *project execution* may start. During execution, particular care must be exercised in supervising critical activities.

- At dates fixed in advance, *cost control* will be carried out. In general, it will then be necessary to update the cost plan and/or to review the time schedule of remaining activities.

- Delays regarding the last date of non-critical events can sometimes be harmless.
  If the retard r of the last date of a non-critical event e does not exceed the smallest independent slack of events immediately following e, the project management can restore the schedule simply by shortening the independent slack of all those follower events by r.

- Whenever the actual execution schedule violates some event slack, time scheduling must be redone — even if the violation took place outside the critical path. In fact, such a violation will delay the project completion if not promptly repaired.

Again, speeding up the activities so as to regain the lost time — when at all possible — will generate additional costs whose amount must be evaluated and whose opportuneness discussed.

4.3 No Choices, No Cycles, Beta-Distributed Durations: PERT

Like CPM, PERT offers methods for designing and controlling large projects. The main difference is that CPM assumes the possibility of determining activity durations as constant values, while PERT takes the uncertainty associated with such estimates explicitly into account.

PERT networks are modified activity networks: the graph structure of the plan representation is the same, but there is no longer a constant duration value associated with each activity. Research and development projects are typical examples of where this approach is often helpful.

PERT is attractive because of the simplicity of the required input data and the clarity of the achieved results. But PERT's easiness is more apparent then real: it rests upon strong assumptions whose acceptability always requires careful checking. Improper use of PERT is not rare, and generally leads to deceptive results. We shall come back to this delicate question.

Like CPM, the PERT technique requires that project plans can be graphically represented as acyclic directed graphs with exactly one source and one sink. Again, the arcs of the graph will represent elementary subprojects.

The crucial difference is that in PERT the duration of each activity a is defined as a *continuous non-negative random variable* T_a of known probability distribution function $F_a(t)$.

In other words, PERT assumes that the probability that the duration of a takes its value in the interval (d_1, d_2) of $\Re^+$ is equal to

$$F_a(d_1) - F_a(d_2) = \int_{d_1}^{d_2} f_a(t)\, dt$$

The function $f_a(t)$ is called the *probability density function* of T_a.

Probability Distribution of Activity Durations

Experience shows that it is quite unrealistic to require persons in charge to express activity durations as probability density functions. PERT overcomes this difficulty in the smooth way we will now illustrate.

PERT assumes all activity durations to be *independent beta distributed random variabless,* and requires for each of them only three — usually easily accessible — estimates:

s   the *shortest* — most optimistic — duration,

n   the *modal* — most likely — duration,

l   the *longest* — most pessimistic — duration.

This is commonly referred to by saying that PERT requires a *three-parameter beta* per activity duration.

Beta distributions well reflect our intuitive understanding of duration of activities. Indeed, the probability density function of a beta distribution:

- is defined on a non-negative interval, and is continuous;
- increases from the shortest duration s up to the modal duration n, and then decreases down to the longest duration l;
- is unimodal — that is, has an unique maximum, in correspondence with the most likely duration n;
- is skewed, so that n can be placed everywhere within interval (s, l);
- has a quite variable shape, and therefore well approximates a wide range of concrete situations.

There are several beta distribution functions for each assignment of s, n and l, and their determination requires in general rather cumbersome calculations. As we will see, this obstacle is easily overcome whenever we consent to constrain the shape flexibility of the beta. This restriction is in practice generally acceptable.

That activity durations are independent random variables having a common distribution function for which mean and variance exist is an essential condition for applying PERT. Indeed, only if this condition holds true is the expected project completion time normally distributed.

The Beta Distribution

We say that a random variable X has a *beta probability distribution* if its normalized density function has the form:

$$
(1) \qquad f(x) = \begin{cases} B(p,\ q)^{-1}\ x^{p-1}\ (1-x)^{q-1} & \text{for}\ \ 0 < x < 1 \\[2mm] 0 & \text{for}\ \ x \le 0\ \ \text{or}\ \ x \ge 1 \end{cases}
$$

where p and q are real strictly positive parameters, and $B(p, q)$ is defined as:

$$
B(p, q) := \int_0^1 x^{p-1}\ (1-x)^{q-1}\ dx\ .
$$

It can be shown [Fi] that $f(x)$ satisfies the requirements of a density function.

The *first two moments of a beta distribution* are [Fi]:

$$
(2) \qquad m_1(x) = p/\,p+q \qquad \text{and} \qquad m_2(x) = p(p+1)/\,[(p+q)\,(p+q+1)]
$$

with p and q as above.

Its *variance* is:

$$
(2') \qquad\qquad \sigma^2(x) = pq\,/\,[\,(p+q)^2\,(p+q+1)]\ .
$$

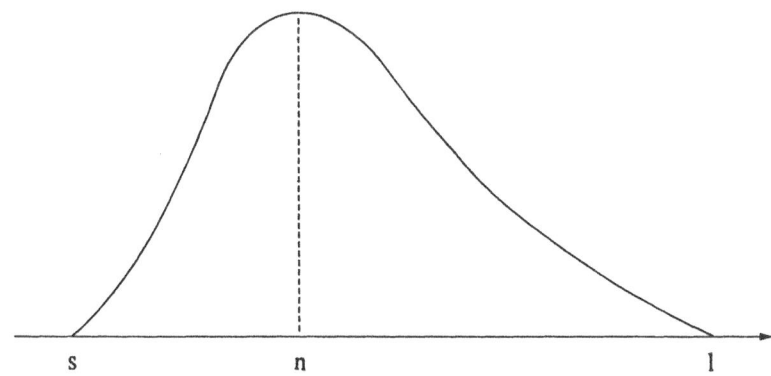

Fig. 4.8

Now, let our project plan be represented by an acyclic directed graph (E, A) with one source and one sink.

Suppose that all activity durations are independent beta distributed random variables. In particular, let us denote by T the random variable expressing the duration t of an activity, and by f(t) the probability density function of T.

Suppose also that the planner has supplied an estimate of s, n and l for each random variable T.

The transformation

$$x = (t - s) / (l - s)$$

will define a random variable X taking values in the interval [0, 1] and whose density function has form (1).

Figure 4.8 shows the diagram of a normalized beta distribution function which is skewed to the right. Values of beta functions can be obtained from the tables of Pearson's curves [Vi].

The standard deviation of unimodal distributions like the beta distribution may be satisfactorily approximated by means of 1/6 of their range [De].

Therefore, the variance of T may be approximated by

(3) $$\sigma^2(t) \approx (l - s)^2 / 36,$$

and the variance of X by

(3') $$\sigma^2(x) \approx 1/36.$$

If we agree to approximate $\sigma^2(x)$ by 1/36, there are two easy ways to find out the values of the parameters p and q characterizing the particular density function of an activity.

Procedure I

We approximate [MR] the expectation m(x) of X by means of the linear function

(4) $$m(x) \approx (4n' + 1)/6$$

where n' is the modal value of X.

Since the modal value n' of X is the transform of the modal value n of T, we get:

(5) $$n' = (n - s) / (l - s).$$

When substituting (5) into (4), the expected value of X becomes:

$$m(x) \approx (4n + 1 - 5s)/[6(l - s)].$$

But the expected value of X is the transform of the expected value of T. It follows that

$$m(x) = [m(t) - s] / (1 - s)$$
$$m(t) = s + (1 - s) m(x)$$

and hence that

(6)
$$m(t) \approx (s + 4n + 1) / 6.$$

For a given estimation of s, n and l, (3) and (6) lead immediately to the variance and expectation of the beta distribution of the random variable T.

Now, which conditions must parameters p and q characterizing f(t) undergo in order to make approximations (3) and (6) acceptable?

The answer is: p and q must be such that the values of $m(t)$ and $\sigma^2(t)$ computed via expressions (3') and (4) equal the values of $m(t)$ and $\sigma^2(t)$ directly obtained from $f_a(t)$.

That is, p and q must be solutions of the equation system obtained by setting equal the two different expressions for $m(t)$ and for $\sigma^2(t)$.

We first transform formulas (4) and (3'), which relate to X, into the corresponding formulas for T:

$$m(t) = s + (1 - s) m(x) = s + (1 - s)p / (p + q)$$
$$\sigma^2(t) = (1 - s)^2 \sigma^2(x) = [(1 - s)^2 pq] / [(p + q)^2 (p + q + 1)] .$$

The equation system for the computation of p and q then becomes:

$$s + (1 - s) p / (p + q) = (s + 4n + 1) / 6$$

(7)
$$(1 - s)^2 pq / [(p + q)^2 (p + q + 1)] = (1 - s)^2 / 36.$$

The right-hand side of the first equation is the transform of the approximation of $m(x)$ assumed in (4). This expression assigns $m(x)$ as a function of n', the modal value of the normalized beta distribution.

n' is, by definition, the only solution of equation $f'(x) = 0$. Solving this equation for x we get

(8)
$$n' = (p - 1) / (p + q - 2).$$

The left-hand side of the first equation of system (7) is also a transformed of $m(x)$ — but obtained by taking $m(x)$ directly in form (2). As a consequence, the first equation of system (7) is equivalent to:

$$p / (p + q) = [4(p - 1) / (p + q - 2) + 1] / 6$$

whose solutions are:

$$p = q ;$$
$$p + q = 6.$$

By substituting into the second equation of system (7), we get, respectively:

(9)
$$p = q = 4 \quad \text{(symmetric density function)} ;$$

(9')
$$p = 3 \pm \sqrt{2} , \qquad q = 3 - \sqrt{2} .$$

These values of p and q are the only values which render approximations (3) and (6) satisfactory with regard to the meaning of n'.

$\square$

(9) and (9') severely constrain the shape of the beta distribution: skewness is limited, and the diagram is so flattened that in a significant interval around the modal value n' the function does not vary significantly.

Procedure I may lead to significant errors for both $m(t)$ and $\sigma^2(t)$. It has been shown [Gr] that the maximum of the absolute error may reach 33% for mean, and 17% for standard deviation. This is the price for the computation facility gained with approximations (3) and (4). If we are not willing to pay this price — for instance because a particular shape of the beta distribution cannot be renounced — we have to turn to

Procedure II

The value of n' is calculated by means of (5), and substituted into (8). The obtained equality is solved for q.

q is substituted into (2'), and the resulting expression of $\sigma^2(x)$ is set equal to 1/36. This substitution is allowed, because assumption (3') still holds true. That way, we get the following algebraic equation of degree three in p:

$$p^3 + p^2(36n'^3 - 36n'^2 + 7n' - 3) + p(-72n'^3 + 52n'^2 - 14n' + 3) + 12n'^3 - 16n'^2 + 7n' - 1 = 0$$

We solve it for p — in general, by numerical approximation — and get q via substitution.

$\square$

This second way of obtaining p and q from estimates of s, n and l has the advantage that it does not constrain the choice of the beta function, even though it is certainly heavier from a computational point of view.

Suppose that our project plan has been represented by means of a PERT network, and that for a certain activity the planner has got the following duration estimates:

$$s = 2, \quad n = 5, \quad l = 14.$$

Which values of parameters p and q characterize beta distributions consistent with these estimates?

First, we try way I.

We agree on approximations (3') and (6), and get so for the mean and standard deviation of the our beta distribution:

$$m = 6, \quad \sigma = 2.$$

Solving system (7), we get:
$$(p = -6916), \quad q = 20748.$$
These values of p and q are not acceptable: they do not match values (9) - (9'), and p is negative. We turn to way II.

We calculate n' using (5), and get
$$n' = 1/4.$$
We then substitute n' into the third degree equation above:
$$16p^3 - 47p^2 + 26p - 1 = 0.$$
The resulting equation has three positive solutions, approximated by:
$$p_1 = 0.679, \quad p_2 = 0.041, \quad p_3 = 2.217.$$
The corresponding values of q are:
$$q_1 = 0.037, \quad (q_2 = -1.877), \quad q_3 = 2.651.$$
(p_1, q_1) and (p_3, q_3) determine two beta distributions consistent with our assignment of s, n, l. The value q_2 is not allowed because it is negative.

The Critical Path Question

PERT scheduling planning is based on expected activity durations, while CPM's is based on known and constant durations.

In PERT, the earliest and latest time of an event e become, respectively, the *earliest expected time* EE and the *latest allowable time* LA. The notion of event slack does not change. The computation of EE and LA is carried out exactly like that of earliest and latest event time in CPM, except for the substitution of expected durations for constant durations.

Also the notion of *critical path* is introduced — again on the basis of expected activity durations — and permits the calculation of the *expected project completion date*. Critical path and expected project completion date are determined as follows.

Consider a PERT network made up of n different paths $p_1, p_2, \ldots, p_n$ from source to sink, and let $D_1, D_2, \ldots, D_n$ be n random variables expressing the execution time of the n paths above. D_i will denote the sum of all random durations of activities on path p_i.

We call a *critical path* every path p_c whose execution time D_c has maximum expectation $m(D_c)$:
$$m(D_c) = \max_i m(D_i).$$

$m(D_c)$ is the *expected project duration*. If, as usual, all project executions are supposed to start at time point zero, $m(D_c)$ is also called the *expected project completion date*.

This way of determining critical paths appears at first glance pleasantly simple. However, its naive use is likely to produce fallacious results.

This is not surprising: only mean activity durations are taken into account. The dispersion of durations around their expected value — that is, the variance of durations — is completely ignored. One of the next sections will be dedicated to the pursuit of this problem.

Now consider a critical path p_i and its random execution time D_i. Since we assumed that activity durations are independent random variables having a common probability distribution for which both mean and variance exist, the central limit theorem can be applied to D_i.

According to this theorem, the sum D_i of the expected durations of activities on path p_i is normally distributed. Obviously, this holds true also for critical paths. The expected project duration D_c is therefore normally distributed too.

Notice that the application of the central limit theorem is based on the tacit assumption that the considered path is made up of sufficiently many activities. A popular rule of thumb requires paths to be made up of at least 15 activities.

For any assignment of values to $D_1, D_2, \ldots, D_n$, we call the *actual project duration* the random variable

$$D = \max_i D_i.$$

For a known property of the expectation of n independent random variables:

$$m(\max_i m(D_i)) \geq \max_i m(D_i)$$

and hence

$$m(D) \geq m(D_c).$$

Therefore, the actual project duration is always greater than or equal to the expected project duration. This is an intrinsic and non-obvious property of PERT time schedules, and it is important to point it out.

Investigation of Project Completion Time

If our PERT network has just one critical path, uncertainty about the project completion date can be appreciated by means of two significant indicators: the standard deviation of expected project completion time and the probability of completing the project not after a certain date.

We have already seen that the central limit theorem allows us to state that the expected project completion time is normally distributed. Therefore:

- the *standard deviation of expected project completion time* is given by
$$\sigma_c = (\Sigma_i \, \sigma_i^2)^{1/2}$$
where $\Sigma_i \, \sigma_i^2$ is the sum of variances of all activities on the only critical path (i varies over the set of critical activities.)

- The *probability of completing the project not after date* d is easily calculated by first computing the ratio
$$z = (d - m_c)/\sigma_c$$
where m_c is the expected value and σ_c the standard deviation of the project completion time — and then taking the area corresponding to z in the tables "Areas under the Normal Curve" [Vi].

Table 4.5

activity	expectation m	standard deviation σ	variance σ²
0→1	13	2	4
0→2	13	3	9
0→3	13	1	1
1→4	8	2	4
2→5	7	1	1
2→6	25	4	16
3→5	12	2	4
3→8	28	3	9
4→6	19	2	4
5→7	19	3	9
5→8	18	5	25
6→9	46	6	36
7→9	20	2	4
8→9	23	2	4

Let us go back to the network in Fig. 4.2, and assume that activity durations are independent random variables whose expectation and standard deviation are listed in Table 4.5.

Since the expectations in Table 4.5 coincide with the constant durations of network 4.2, we have just one critical path, the same one outlined in the figure. (Critical activities are underlined in Table 4.5 too.)

The expected project completion time obviously coincides with the project completion time of the CPM network, and is

$$m_c = 13 + 8 + 19 + 46 \approx 86 \text{ time units.}$$

The standard deviation of expected project completion time is

$$\sigma_c = (\Sigma_i \sigma_i^2)^{1/2} = (4 + 4 + 4 + 36)^{1/2} = (48)^{1/2} \approx 6.93 \quad \text{time units.}$$

We have

$$z = (100 - 86) / 6.93 = 2.02$$

which gives 0.97831 for the probability of completing the project not after 100 time units.

In Fig. 4.9 the (normal) distribution of the expected project completion time is shown. The area representing the probability of completing the project on or before date 100 is shadowed.

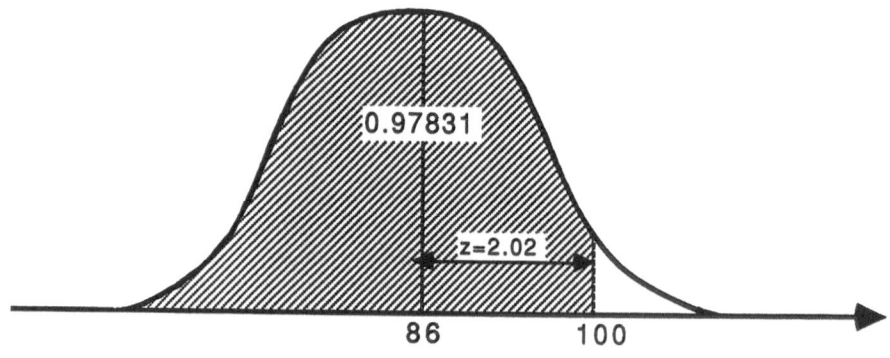

expected time

Fig. 4.9

Warnings about Improper Use of PERT

Letting activity durations be random variables assigns to each source-sink path in the network a non-zero probability to be the path with longest duration, or execution time — that is, to be the critical path.

PERT networks have by definition an infinite number of possible executions, each one determined by associating a particular duration value to every activity in the network, in accordance with the activity-specific distribution function.

In every execution there is at least one source-sink path with longest execution time. Every random variable D_i expressing the execution time of a source-sink path p_i has a certain probability to take maximum value.

Consider a PERT network made up of n different paths $p_1, p_2, ..., p_n$ from source to sink. All of them have a certain probability to be critical, that is, to be the path with longest execution time. Since there are infinitely many different executions, the probability for path p_i to be critical can be estimated by means of the relative frequency, in a sufficiently large number of simulation runs, of the executions in which p_i had maximum execution time.

When the network is made up of many different source-sink paths, it may happen that paths with the greatest probability of being critical nevertheless have, in absolute terms, a very low probability to be critical. And, of course, the paths with the greatest probability to be critical are not necessarily those with the maximum expected duration.

Since PERT time scheduling relies significantly on the notion of critical path, the circumstances of PERT's use must be attentively examined with reference to the considered application. Choosing critical paths without the necessary care is likely to have undesired consequences.

The mini-network in Fig. 4.10 will illustrate the former considerations. Expected activity durations are inscribed on corresponding arcs.

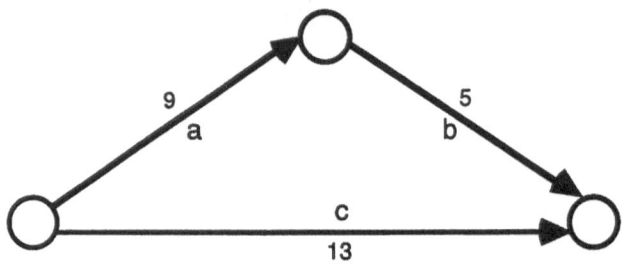

Fig. 4.10

In network 4.10 the critical path determined by means of expected activity durations is made up of activities a and b: its expected execution time is 14, while the expected execution time of the other source-sink path is 13.

Suppose that in a sufficiently large number of simulation runs, carried out according to the beta functions of activities, path ab has had the maximum execution time with frequency 0.37, path c has had the maximum execution time with frequency 0.56, and the two paths have had equal duration with frequency 0.07.

If the frequencies above are taken as probability estimates, the critical path determined by means of expected durations is not the path with the highest probability of being the path with maximum execution time. And its probability to be the path with maximum duration is anyhow too small to give any confidence.

A further problem is that activities having the maximum probability to belong to a path with maximum execution time may not belong to a path with highest probability of having the maximum execution time.

We will clarify this point using the small network in Fig. 4.11. Suppose we carry out a congruous number of simulation runs assigning a random duration to each activity in accordance to its beta function. Suppose further that we collect the following frequencies:

path ab	had the longest duration in	20% of the runs
path ac	had the longest duration in	20% of the runs
path ad	had the longest duration in	20% of the runs
path ef	had the longest duration in	40% of the runs

The frequency with which an activity was on the path with longest duration can be immediately deduced. We have inscribed such frequencies on the corresponding arcs in Fig. 4.11.

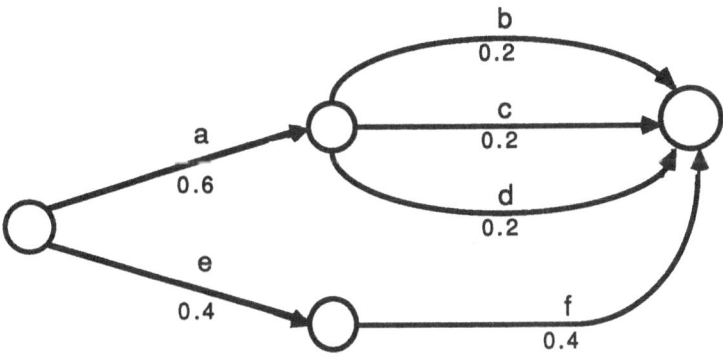

Fig. 4.11

Using again the frequencies above for estimating probabilities, we see that activity a, which has probability 0.6 of belonging to a path with maximum execution time, does not belong to path ef, which is the path with the highest probability of having the maximum execution time.

Such considerations seem to suggest giving up the notion of critical path, in favor of some better suited criticality indicator focused on single activities. The activities to keep under tight control would then no longer be those whose expected delay equals zero, but those having the highest degree of criticality. In practice the probability of belonging to the path with longest expected execution time is used as criticality indicator. This probability is, on its turn, estimated by means of frequencies.

These problems are sometimes avoided by skipping from PERT to Monte Carlo methods. This is done in the following way:
- we determine random executions of the considered PERT network by assigning a random duration to each activity according to its beta function;
- we provide a sufficiently large set of such random executions of our network;
- we apply the CPM algorithms to every one of these random executions;
- we store the outcomes of all CPM calculations, and then work them out by means of usual statistical methods.

The difficulties we have illustrated are mitigated in case the network is made up so that no two source-sink paths share an activity.

Let D_c and D_p be two random variables expressing, respectively, the expected duration of a critical path c and the expected duration of another generic source-sink path p. In our hypotheses, D_c and D_p are independent normally distributed random variables. It follows that the random variable $(D_c - D_p)$ has a gaussian distribution too, and that

$$m(D_c - D_p) = m(D_c) - m(D_p) > 0$$
$$\sigma^2(D_c - D_p) = \sigma^2(D_c) + \sigma^2(D_p).$$

Hence

$$P\ [D_c - D_p > 0] > 0.5.$$

If all source-sink paths are independent, the probability that the critical path has longer duration than a generic source-sink path is greater than 0.5.

Unfortunately, this weak result usually has to be renounced too: independence of all source-sink paths is very seldom a property of realistic networks.

Last, we will discuss the question of the statements about the required project completion date.

Let D_c and D_p be random variables expressing, respectively, the expected duration m_c of a critical path c and the expected duration m_p of a generic source-sink path p. By definition, m_c is greater than m_p.

Figure 4.12 shows that if m_c is not much greater than m_p but the variance of D_p is significantly greater than the variance of D_c, the execution time of path p is more likely to exceed the requested project completion date ρ than the execution time of the critical path.

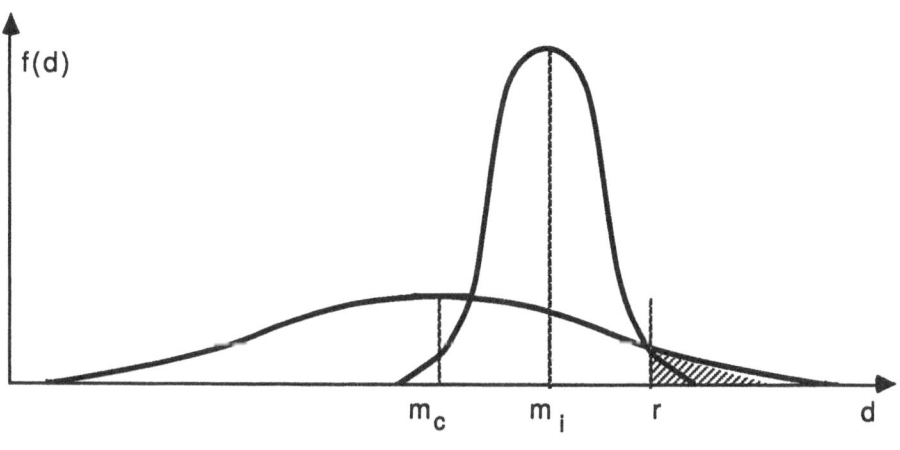

Fig. 4.12

PERT Cost Planning and Optimization

CPM cost planning and optimization algorithms can be applied in a PERT environment too. PERT networks may indeed be regarded as CPM diagrams simply by interpreting expected activity durations as known constant durations.

If handling of cost uncertainty is required — which is surely in the spirit of PERT technique — it can be treated like duration uncertainty. In particular:

- to each activity a pessimistic, an optimistic and a modal — or most likely — cost is associated;
- activity costs are assumed to be stochastically independent random variables;
- every activity cost is defined to be a continuous beta distributed random variable.
- The planned project cost is defined to be the random variable
$$C = \Sigma_i C_i$$
where C_i is the expected cost of activity i and the sum is over *all* project activities.

If the network is made up of a sufficiently large number of activities, the central limit theorem allows us to state that C is normally distributed and that its mean and variance are:
$$m (C) = \Sigma_i m(C_i),$$
$$\sigma^2(C) = \Sigma_i \sigma^2 (C_i).$$

The assumption that activity costs are stochastically independent is not always realistic. It is however generally accepted because the investigation of cost interdependencies is generally too complex to be undertaken.

Operational Considerations

PERT is suitable for planning and controlling projects which are made of activities with highly uncertain completion time. The planner must be able to fix the shortest, the longest and the most likely duration for each activity. As in CPM, choices and cycles cannot be represented.

The successful use of PERT requires previous, accurate examination of the project features. We have seen how an improper use of this apparently simple technique may lead to bitter deceptions.

The following is a memorandum of conditions which must be be verified before deciding to apply PERT.

<u>Memorandum</u>

In order to apply PERT:
-   It must be reasonable to assume that activity durations are independent and beta distributed random variables. In particular, we must check if their distribution can be assumed to be unimodal.
-   Either we are prepared to deal with cumbersome computations for deducing the parameter values of the betas from the three duration estimates, or we agree to constrain the shape of the beta function associated with the activity durations, both in its skewness and variability around the modal value.
-   We are conscious of the consequences of the circumstance that the dispersion of activity durations around the expected value does not play any role in the determination of critical paths.
-   We are aware that the application of the central limit theorem for getting expectation and variance of the project completion date requires that critical paths are made up of a sufficiently large number of activities.
-   We remember that the project execution time is expected to be greater than or equal to the planned project duration.
-   We know that each source-sink path has a certain probability to be a critical path, and that if the network is made up of many different source-sink paths, the source-sink

path with the greatest probability to have the longest execution time may have a very low probability of being a critical path.

- We remember that critical paths determined on the basis of expected activity durations may not have maximum probability of being the path with longest execution time. This probability has anyway to be regarded as a significant confidence indicator.
- We are aware that activities having maximum probability of belonging to a path with longest execution time may not belong to a path with the highest probability of being critical.
- Whenever considering the required project completion date ρ, we will remember that the completion date of a generic path may be more likely to exceed ρ than that of a critical path.

So far it should be clear that there must be strong motivation for preferring PERT to CPM, motivation which will in general refer to the presence of a high degree of uncertainty in activity durations or in costs. Once we have decided to use PERT, its application protocol will follow the guidelines indicated for CPM except for obvious differences.

4.4 Choices, Cycles, Random Durations: GERT

Like CPM and PERT, the planning technique GERT is based on the digraph representation of project plans. Its representation language, too, is a variant of activity networks, a variant arising from quite radical modifications, both in network structure and interpretation.

As we will see, augmented representation capability is obtained at the price of a reduction in the computational possibilities. Whenever the full representation power of GERT is exploited, simulation will turn out to be the only network analysis method.

We will introduce the GERT plan representation language by contrasting it with both the CPM and PERT representation languages.

Event Realization "Logics"

A *GERT network* is a directed graph. Like in CPM and PERT, arcs represent activities. More sources, more sinks and cycles are allowed, and vertices of different types can be specified. Here we will consider only GERT networks with one source.

As in CPM and PERT, each pair $e = (IN, OUT)$ of *maximal* not both empty sets of activities such that

$$\forall\, a \in IN \quad \forall\, b \in OUT : \quad a < b \quad \wedge \quad \neg\, (\exists\, c \in A : \quad a < c < b)$$

is interpreted as a time point τ, and called an *event* or *milestone*.

We say that *the event* e *has been realized at time point* τ.

In GERT a "logic" is explicitly assigned to each event. This logic specifies how the time point τ at which event e is realized relates to both the *input side* IN and the *output side* OUT of e.

Various event logics are allowed in GERT networks. They are best understood by considering for each activity belonging to set IN the time point at which it terminates, and for each activity of set OUT the time point at which it starts.

The *input side* IN of a GERT vertex may be declared to be of three different types:

- *AND type*. This is the logic of CPM and PERT networks: an event e is realized at the first time point all its IN activities have terminated. Time point τ is the time point at which the last input activity or activities of e terminate(s).

- *Inclusive-Or (IOR) type*. An event e is realized at the first time point one or several of its IN activities terminate. τ is the first time point at which an input activity of e terminates.

- *Exclusive-Or (EOR) type*. An event e is realized at the first time point exactly one of its input activities completes. τ is the time point at which one — but not more than one — input activity of e first terminates. If several activities complete simultaneously, e is not realized.

Figure 4.13 shows the graphical representation of the three different input side types of a GERT event.

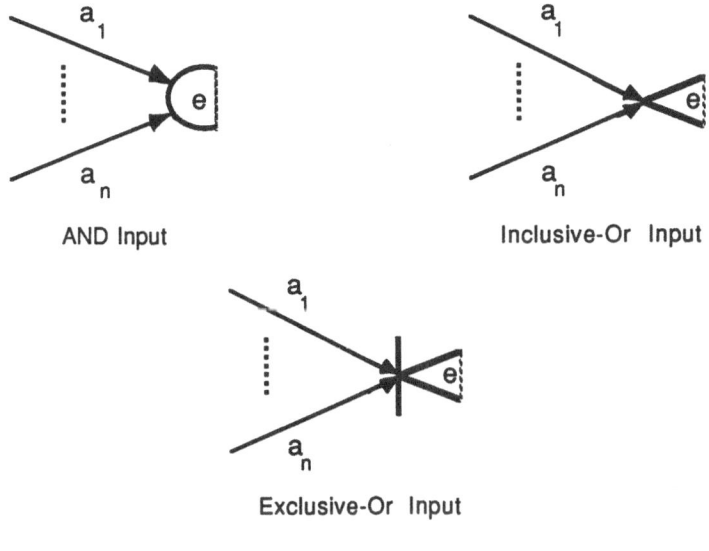

AND Input   Inclusive-Or Input

Exclusive-Or Input

Fig. 4.13

The *output side*   OUT of a GERT vertex may be declared to be of two different
types:
-   *Deterministic type.* This is the logic of CPM and PERT networks: *all* OUT activities
    of an event e start immediately at time point τ at which e is realized. With certainty.
-   *Stochastic type.* With each of activities $a_1, a_2, ... , a_n$ belonging to the OUT set of
    an event e, a mapping p_i of $\Re^+$ into the interval $[0,1]$ is associated.

    For every $\tau \in \Re^+$ the value $p_i(\tau)$ represents the *conditional probability* that activ-
    ity a_i begins at time point τ given that its initial event was realized at time point τ. Of
    course, for every $\tau \in \Re^+$, $\sum_i p_i = 1$.

    At time point τ at which e is realized, exactly one OUT activity starts. Activity a_i has
    probability $p_i(\tau)$ to be the starting activity.

    If mappings p_i are constant functions — that is, if for all $\tau \in \Re^+$, value $p_i(\tau)$
    equals p°_i — the probability associated with activities is no longer conditional. Like
    in PERT.

Figure 4.14 shows the graphical representation of the two different output side types
of a GERT event.

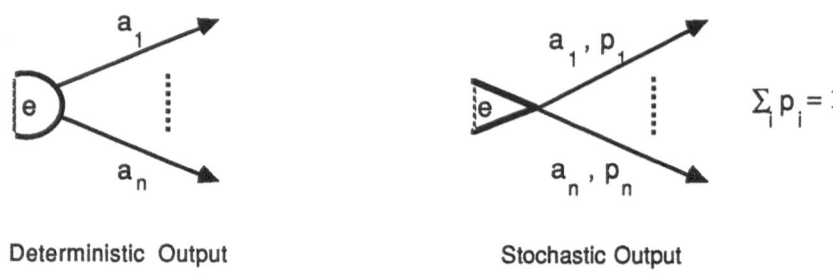

Deterministic Output   Stochastic Output

Fig. 4.14

By combining the three input types with the two output types we get six different kinds of events. Their graphical representation is shown in Fig. 4.15.

In GERT terms, we might say that the vertices of CPM and PERT networks are all deterministic AND vertices.

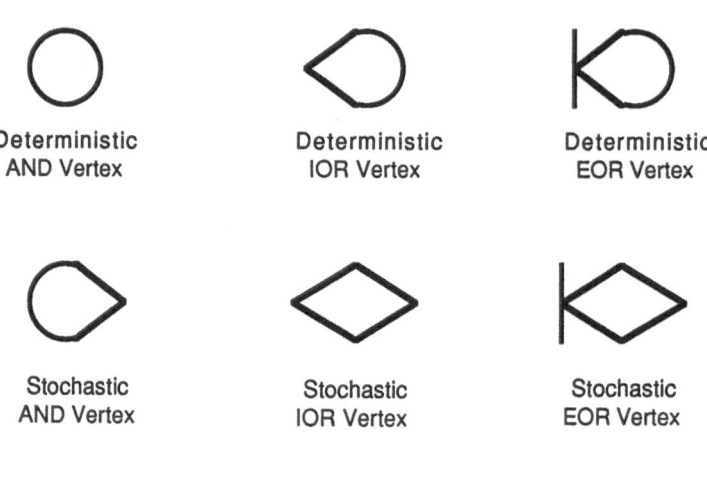

Deterministic   Deterministic   Deterministic
AND Vertex   IOR Vertex   EOR Vertex

Stochastic   Stochastic   Stochastic
AND Vertex   IOR Vertex   EOR Vertex

Fig. 4.15

Also, the GERT technique allows the specification of AND/AND events whose realization takes place at the r-th completion of an input activity.

The natural number r is a characteristic of each event, and may be assigned differently for the first and for all subsequent event realizations. The values of r are inscribed in network vertices, both for the first and the subsequent event realizations.

The r completions of input activities which are necessary for event realization may be counted in any order or combination, even allowing for repetitions.

An example is shown in Fig. 4.16. The first realization of event e requires the termination of three input activities, in any order or combination. Subsequent realizations of e require infinitely many completions of input activities. This last specification tells that a second realization of event e must be considered impossible in practice.

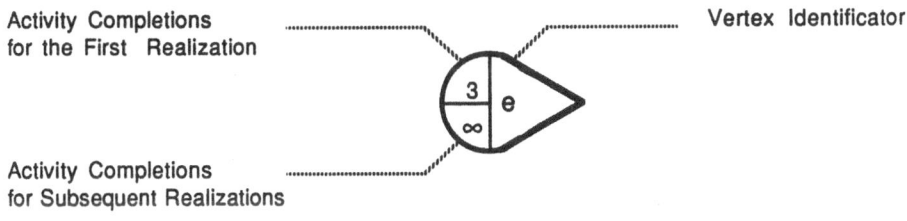

Fig. 4.16

Network Structure

The definition of GERT networks requires — similarly as for CPM and PERT — the existence at least of an initial event representing plan beginning and called *the source* . Plan completion may be represented — unlike in CPM and PERT — by means of different terminal events called *sinks*.

Here we assume that the *source is unique*, and that it will be activated at time point zero. If there are several sources, we will adjoin a "new" vertex to the network, and connect it to every source vertex by means of a dummy activity. This will give a unique source.

GERT requires plan representations to have *at least one sink,* and all sinks to be reachable from the source. Different sinks are good for representing plans with several possible terminations, for instance, plans which may end up either with success or with abortion.

Cycles are allowed in GERT networks. Activities can be specified to be executed repeatedly, for a given number of times. Thus, iterated execution of processes may be represented.

Probabilistic branching is provided. This allows the representation of choices governed by probability laws. By convention, probabilities assigned to activities — to arcs — must be greater than zero.

Branching is necessary for giving sense to the co-presence of several sinks. If all events were of deterministic AND/AND type — as is the case in standard activity networks — all sinks would by plan be realized ultimately. Plans with mutually exclusive terminations could not be represented.

Last, we assume that:

- the duration of the n-th iteration of an activity may depend on its starting time, but not on the past history of the network execution;
- the output activity starting at the realization of a STEOR vertex may depend on the time point, but not on the past history of the network execution.

Probability Distribution of Activity Durations

We have seen that CPM requires that the durations assigned to activities are constant values, while in PERT durations are supposed to be random variables with a three-parameter beta distribution.

GERT, too, assumes activity durations to be random variables, but offers a rich choice of time distributions.

For each activity a the planner is required to assign a *conditional duration distribution function* F_a defined as

$$F_a(d \mid \tau) := \begin{cases} P\ (D_a \le d \mid \text{execution of a begun at time } \tau) & \text{for} \quad d \ge 0 \\ 0 & \text{for} \quad d < 0 \end{cases}$$

where D_a is the random variable representing the duration of activity a.

For each activity the planner is required to assign the type of the conditional distribution function as well as the corresponding parameter values. As a special case, distribution functions F_a may be assigned independently of τ, that is non-conditionally.

Table 4.6 displays a choice of typical GERT distribution types.

Table 4.6

Code	Distribution
1	constant value
2	normal distribution
3	uniform distribution
4	Erlang distribution
5	lognormal distribution
6	Poisson distribution
7	gamma distribution
8	beta distribution
9	three-parameter beta distribution

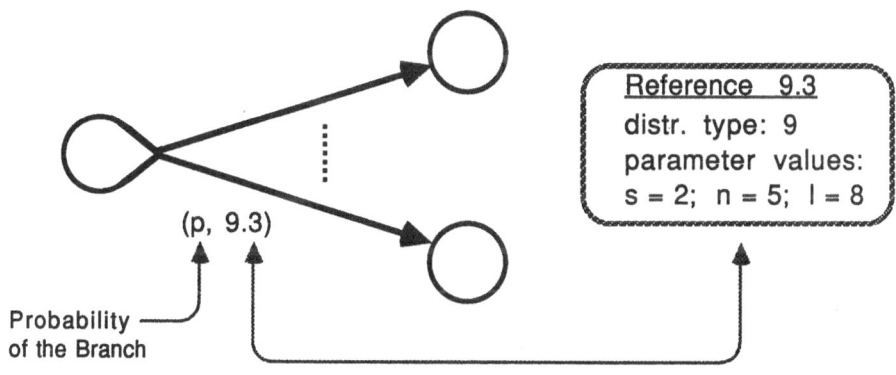

Fig. 4.17

Distribution type and parameter values can be referenced in the arc inscription of activities as shown in Fig. 4.17. The list of allowed probability distributions of activity durations can be extended if necessary.

STEOR Networks and Markov Renewal Processes

STEOR networks are GERT networks whose vertices are all of Stochastic Exclusive-Or (STEOR) type. They deserve particular attention because they allow one to find out the expected number $\rho_{ij}(\tau)$ of occurrences of an event j within the τ time units following the occurrence of another event i. In particular, the *activation distribution of the sinks* and *unconditional distribution function of the project duration* can be calculated.

These results depend on the possibility of associating a homogeneous Markov renewal process to every network event, and can be obtained only for STEOR networks. As the procedure is rather cumbersome, we will limit ourselves to a brief sketch. Interested readers will find a detailed presentation of the subject in [NS].

Let G be a STEOR network, and $\{0, 1, \ldots, n\}$ the event set of G. For simplicity in our exposition, two further hypotheses on G will be appropriate — hypotheses not implying any loss of generality:
- We assume that G is be *free of parallel arcs*.
  If this is not the case, every pair of parallel arcs $(i \rightarrow j)'$ and $(i \rightarrow j)''$ will be replaced by a unique arc $i \rightarrow j$ whose conditional execution probability p and conditional distribution function F are given by
  $$p = p' + p'' \quad \text{and} \quad F = (p'F' + p''F'')/(p' + p'')$$
  where p' and p'' denote the conditional probabilities of $(i \rightarrow j)'$ and $(i \rightarrow j)''$, and F' the conditional distribution functions of $(i \rightarrow j)'$, and F'' that of $(i \rightarrow j)''$.
- We exclude the presence of cycles made up just of dummy activities.
- We postulate that subsequents realizations of events connected by a dummy activity are separated by arbitrarily small time intervals.

We denote with p_{ij} the conditional execution probability of activity $i \rightarrow j$ — the activity starting at event i and ending at event j — and with F_{ij} the conditional duration distribution function of $i \rightarrow j$.

Recall that, as all vertices of G are of STEOR type, $p_{ij}(\tau)$ is the *conditional probability* that activity $i \rightarrow j$ begins at time point τ given that event i was realized at time point τ.

$F_{ij}(d \mid \tau)$ is the probability that the duration of $i \rightarrow j$ is less than or equal to d, again given that event i was realized at time τ.

At each time point τ, at most one event of G can be realized. In fact, G has a unique source, activated at time point zero. Since all vertices are STEOR, exactly one output activity starts when an event is realized. At any time point τ, at most one activity can terminate, and therefore at most one event can occur. Event realizations will follow one after the other.

Let X_r indicate the r-th event which occurs, τ_r the time point of the realization of X_r, and τ_0 the source activation time:

$$\tau_0 < \tau_1 < \tau_2 < \ ..... \ .$$

An activity $i \to j$ may be seen as a transition from state i to state j, a transition with conditional execution probability p_{ij} and with conditional duration distribution function F_{ij}. Our hypotheses ensure that

$$P \ (X_{r+1} = i \ \wedge \ \tau_{r+1} - \tau_r \le \tau \ | \ X_r, \dots, X_0; \ \tau_r, \dots, \tau_0) =$$
$$= P \ (X_{r+1} = i \ \wedge \ \tau_{r+1} - \tau_r \le \tau \ | \ X_r)$$

for every natural number r, every event i and every positive real number τ.

Synthetically, this can be expressed by saying that $(X_r, \tau_r)_{r \in \mathfrak{z}+}$ is a *Markov renewal process* with state space $E \times \mathfrak{R}^+$, where $E = \{0, 1, \dots, n\}$ is the *set of events* of **G**. Event realizations are often called *renewals* in this context.

The *transition matrix* of the above Markov renewal process is the $(n+1) \times (n+1)$ matrix $[p_{ij}]$. The *transition functions* of the process are :

$$Q_{ij}(\tau) \ := \ \begin{cases} P \ (X_{r+1} = j \ \wedge \ \tau_{r+1} - \tau_r \le \tau \ | \ X_r = i) & \text{for} \quad \tau \ge 0 \\[2mm] 0 & \text{for} \quad \tau < 0 \end{cases}$$

Since the functions $Q_{ij}(\tau)$ do not depend on r, we say that the Markov renewal process $(X_r, \tau_r)_{r \in \mathfrak{z}+}$ is *homogeneous*.

We have:

$$Q_{ij}(\tau) = p_{ij} \ F_{ij}(\tau) ,$$

It can be proved [NS] that the first component of the Markov renewal process above — that is, $(X_r)_{r \in \mathfrak{z}+}$ — is a homogeneous Markov chain whose state space is characterized by the following properties:

- If event i is not a sink and does not belong to any cycle, then the probability that after a realization, i will occur again, is less than one. In other words, non-sink events outside of cycles represent transient states.

- If event i belongs to a cycle C, then the probability that after a realization of i, a non-sink event j not belonging to C will occur, is greater than zero. That is, events belonging to cycles never represent persistent states.

- With probability zero, after the realization of a sink event — and only after the realization of a sink event — any event may occur. That is, sinks and only sinks represent absorbing states.
- With probability one, exactly one sink will be realized during each network execution. Again with probability one, this sink will be realized within a finite time from the occurrence of the source event.

The MRP Method

We are now in a position to illustrate the MRP (Markov Renewal Process) method for finding out the activation distribution of the sinks and the unconditional distribution function of the project duration of a STEOR network.

We have seen that — with certainty — exactly one sink of **G** will be realized during each network execution. The activation distribution of the sinks and the unconditional distribution function of the project duration can be calculated by means of the so-called renewal functions. We will first introduce them.

For any two events **i** and **j** and any non-negative real number τ, let $\#_{ij}(\tau)$ be the count of renewals (realizations) of **j** within the time interval $[0, \tau]$ whose time point 0 coincides with the occurrence of **i**.

If $E[\#_{ij}(\tau)]$ denotes the expectation of $\#_{ij}(\tau)$, we call the function $\rho_{ij}(\tau)$ defined by

$$
\rho_{ij}(\tau) := \begin{cases} E[\#_{ij}(\tau)] & \text{for } \tau \geq 0 \\ 0 & \text{for } \tau < 0 \end{cases}
$$

the *renewal function* of process $(X_r, \tau_r)_{r \in \mathfrak{J}+}$.

$\rho_{ij}(\tau)$ gives the expected number of renewals of **j** in the interval $[0, \tau]$ specified as above.

The renewal functions $\rho_{ij}(\tau)$ can be computed with the help of the so-called *r-step transition functions* $Q_{ij}^r(\tau)$. These are defined as

$$
Q_{ij}^r(\tau) := \begin{cases} P(X_r = j \wedge \tau_r \leq \tau \mid X_0 = i) & \text{for } \tau \geq 0 \\ 0 & \text{for } \tau < 0 \end{cases}
$$

where, as above, r is the index of the renewal process, and i and j vary over the set of event labels $\{0, 1, ... , n\}$.

It can be proved that

$$\forall \, i, j \in E : \quad \rho_{ij}(\tau) = \Sigma_{r = 0, 1, ...} \; Q_{ij}^{r}(\tau) .$$

For the r-step transition functions $Q_{ij}^{r}(\tau)$ we have:

$$Q_{ij}^{1}(\tau) = Q_{ij}(\tau)$$

$$Q_{ij}^{r}(\tau) = \Sigma_{k \in E} \; Q_{ik}^{r-1} * Q_{kj}^{1}(\tau) \quad \text{for } r > 1$$

where " * " denotes the convolution of the two functions. Recall that

$$Q_{ik}^{r-1} * Q_{kj}^{1}(\tau) =: \quad \begin{vmatrix} \int_{0}^{\tau} Q_{ik}^{r}(\tau - x) * Q_{kj}^{r}(dx) & \text{for } \tau \geq 0 \\ \\ 0 & \text{for } \tau < 0 \end{vmatrix}$$

We call the set of the renewal functions $\rho_{0s}(\tau)$ — where 0 identifies the source and s any sink — the *activation distribution of the sinks of* G.

Since every sink is realized at most once per network execution, the probability that sink s is activated not later than time point τ is:

$$\alpha_{s}(\tau) = P \, [\, t_{s} \leq \tau \,]$$

where t_{s} is the time point of the occurrence of s. The probability that s will ever be activated is:

$$a_{s} = P \, [\, t_{s} \leq + \infty \,] = \lim_{\tau \to + \infty} \alpha_{s}(\tau).$$

The *conditional probability* that s is realized not later than time τ, given that it takes place at all, is:

$$\delta_{s}(\tau) = P \, [\, t_{s} \leq \tau \,|\, t_{s} < \infty \,];$$

and for $a_{s} > 0$:

$$\delta_{s}(\tau) = \alpha_{s}(\tau) / a_{s} .$$

The *unconditional distribution function of the project duration* is:
$$\delta(\tau) = \Sigma_{s} \alpha_{s}(\tau) = \Sigma_{s} a_{s} \delta_{s}(\tau).$$

Consider the STEOR network in Fig. 4.18. The first entry in the tuples on the arcs is the constant duration of the corresponding activity; the second entry is the probability that this activity will start given that its initial event is realized. We shall compute:
-   the probabilities a_{7} and a_{8} that the sinks will be realized;

- the conditional distribution functions $\delta_7(\tau)$ and $\delta_8(\tau)$ of project duration given that either sink is realized;
- the unconditional project duration distribution function $\delta(\tau)$.

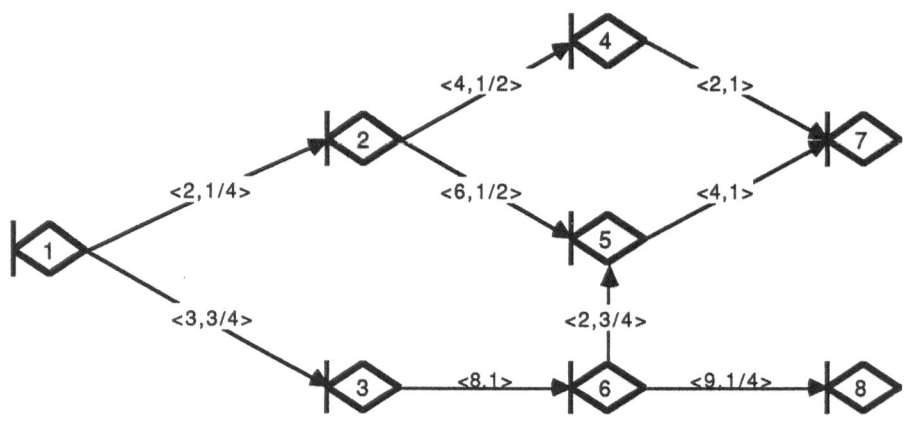

Fig. 4.18

For the four source-sink paths of the network, we get:

Path	Realization-Probability	Duration
1247	1/8	8
1257	1/8	12
13657	9/16	17
1368	3/16	20

and hence:

$a_7 = 1/8 + 1/8 + 9/16 = 13/16;$ $a_8 = 3/16;$

$\alpha_7(8) = 1/8;$ $\alpha_7(12) = 1/8 + 1/8 = 1/4;$ $\alpha_7(17) = 1/8 + 1/8 + 9/16 = 13/16;$

$\alpha_8(20) = 3/16.$

The conditional distribution function $\delta_7(\tau)$ must be computed at the time points 8, 12 and 17, at which sink 7 may occur. The conditional distribution function $\delta_8(\tau)$ must be computed at time point 20, the only one in which sink 8 may be realized. We get:

$\delta_7(8) = (1/8) : (13/16) = 2/13; \quad \delta_7(12) = \alpha_7(12) : (13/16) = 4/13;$
$\delta_7(17) = \alpha_7(17) : (13/16) = 1;$
$\delta_8(20) = \alpha_8(20) : (3/16) = 1.$

Figures 4.19 and 4.20 show the diagrams of the conditional distribution functions $\delta_7(\tau)$ and $\delta_8(\tau)$.

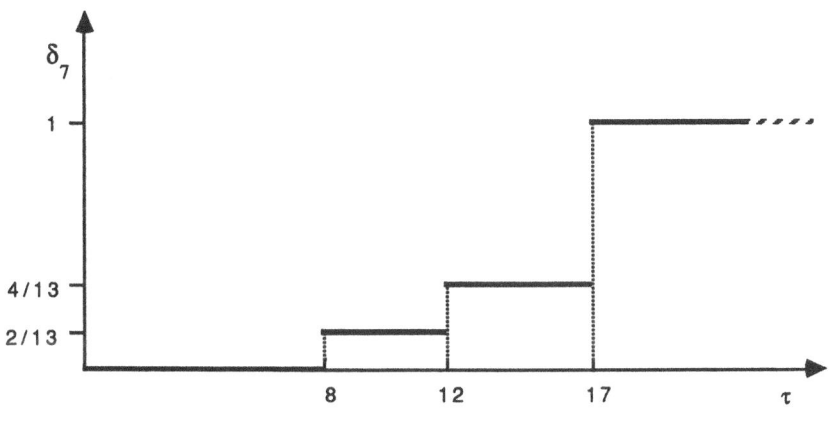

Fig. 4.19

The unconditional project duration distribution function $\delta(\tau)$ is:

$$\delta(\tau) = a_7 \, \delta_7(\tau) + a_8 \, \delta_8(\tau) = (13/16) \, \delta_7(\tau) + (3/16) \, \delta_8(\tau).$$

The jump points of $\delta(\tau)$ are $\tau = 8, 12, 17, 20,$ and we get:

$\delta(8) \quad = (13/16) \cdot (2/13) + (3/16) \cdot 0 \ = 1/8$
$\delta(12) = (13/16) \cdot (4/13) + (3/16) \cdot 0 \ = 1/4$
$\delta(17) = (13/16) \cdot 1 + (3/16) \cdot 0 \ = 13/16$
$\delta(20) = (13/16) \cdot 1 + (3/16) \cdot 1 \ = 16/16 \ = 1$

The diagram of $\delta(\tau)$ is shown in Fig. 4.21.

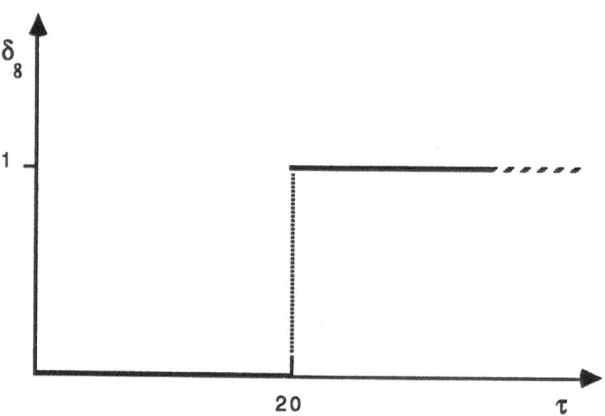

Fig. 4.20

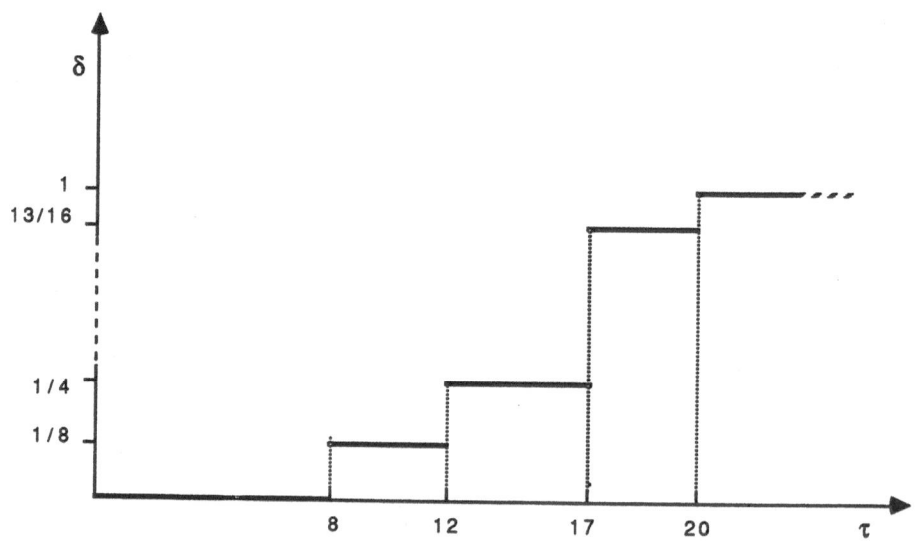

Fig. 4.21

Simulation of GERT Networks

Project plans are often too complex to be investigated analytically: sometimes there are no known algorithms which solve our problems, and direct experimentation is usually not practicable. However, information about the expected execution of a project plan can always be gathered via simulation.

Simulation techniques are based on repeated running of suitable models of the systems to be investigated. Sample observations are collected and statistically worked out. The outcome will be sample statistics about some system performance indicators.

Probabilistic system models are called stochastic when the uncertainty involves time. The simulation of probabilistic and particularly of stochastic system models is often based on the so-called *Monte Carlo method* : values to be assigned to random variables in simulation runs are chosen at random but so that over a large number of runs their frequency distribution approximates a specified probability distribution function.

The analytical investigation of GERT networks is — as we have seen — not practicable as a rule. Even if we are dealing with STEOR networks, the network size is likely to make an analytical solution too hard. This is the price for GERT's structural flexibility and for the rich choice of distribution functions.

GERT is naturally oriented toward simulation, and the popularity of this technique is to a significant extent due to availability of good GERT simulation packages [BP, CM1, CM2].

In the previous sections we have seen how duration of activities is represented in GERT networks. Costs of activities, resources requested to perform a task (customer to be served in a queue, etc.) can also be represented similarly. As a consequence, there are packages dedicated to cost planning, to resource allocation questions, to queuing problems, etc. A careful presentation of the general purpose GERT simulation package GERTZ III Z supplied with numerous real-world application examples can be found in [CM2].

Here we will only describe some standard features of GERT simulators dedicated to time scheduling.

The simulator inputs a GERT network together with information about statistics to be collected. The network must be equipped with the necessary information over activities duration. In particular, the following data must be input:

- for each vertex:
  the type of input and output;

the number of activity completions required for one node realization — separately for the first and the subsequent realizations;

* for each arc:

  its starting probability given that its initial event was realized;

  the sort of duration, constant or random;

  the probability distribution function of the duration — if the latter has been declared to be random;

  the parameter values of the distribution function.

Information about collection of statistics will be input by declaring that some network vertices belong to one or more of following types:

SOURCE	one single vertex allowed
SINK	more than one vertex allowed
STATISTIC	statistics about vertex realization time collected
MARK	reference node for a statistic vertex.

Statistics will only be collected at statistic vertices and sinks. Source and mark nodes are just reference points for data sampling.

In GERT simulators dedicated to time scheduling at each simulation run, the following data will be collected both at statistic vertices and sinks:

- the date of first realization;
- all dates of subsequent realizations;
- the time interval between subsequent realizations of the same vertex;
- the delay between first termination of an input activity and realization of the vertex;
- the delay between realization of the vertex and realization of a specified mark node.

5. Clock-Independent Planning: Petri Nets

Up to now project plans have been represented as networks of activities — activities specified to take a certain amount of time and/or to incur a certain cost, and which are assumed to start as soon as a specified set of prior activities terminate. Various modes of representing activity attributes were introduced in the plan representations discussed so far. In all cases a single cause for activity start was taken into account: the termination of some set of prior activities.

We may say that, in activity networks, activities lying on the same directed path are seen to be *causally dependent,* while those which do not are assumed causally independent — or *concurrent.* Causal dependence entails time precedence. Causal independence is assumed equivalent to a lack of prescribed temporal ordering.

Activity networks assume the existence of a global clock to which planned durations refer and by which actual durations are measured. This harmonizes with the circumstance that activity networks are used to plan human coordinated work, work which refers to at least one globally valid clock: the succession of days and nights.

While these assumptions may appear reasonable enough, they fail to describe many important aspects of causal interconnection between activities in project plans. Activities have several causes and affect other activities in several ways. In many cases, activities must wait on other activities because of specific resources which they share.

Project plans expressed in Petri net form are systems of causally interconnected state and state transition components. This provides the possibility to model various types of interdependence based on shared resources, and also to consider the relationship between global states and local state components.

Petri nets allow — like GERT networks — the representation of projects with cycles, choices, and multiple beginnings and endings. In addition, states of plan execution can be represented explicitly, together with the transitions between them.

The notion of plan execution state implies that the elementary components of execution states are related to a unique reference entity, a global supervisor. At any given point in time, it is the supervisor who — conceptually — forms elementary state components into an actual plan execution state.

Repeated executions of the same plan will differ in the duration of its component parts, and one can expect that the sequence of global states as seen by the plan supervisor will differ from one execution to the next. In the Petri net context, the sequence of execution states replaces the global clock assumed for activity networks. Petri nets are therefore mainly used for the clock-independent analysis of plans.

Extensions of Petri nets which take account of global clock ticks have been proposed, both deterministic and stochastic. Their application is very often based on simulation, because the analysis capabilities of such models are restricted to quite special cases. Their applicability to project design and control is not yet clear, however, so we do not introduce them in this book.

Planning with ordinary Petri nets while taking into account times, and hence costs, is indeed already possible. This grafting will allow us to exploit Petri nets for project engineering. By doing this we can develop not only a finer understanding of the causal interconnections among the planned activities, but also a unitary representation of plan structure and plan dynamics, and the monitoring of resource flows and plan executions.

Petri nets are also suited for the disciplined development style introduced in Chapter 3, and support strategic decision making over alternative plan executions.

Petri nets were developed from the early work of C. A. Petri in order to represent and analyze general systems from the viewpoint of causal interconnections among the elementary system components.

There are several types — or classes — of Petri nets, suited for plan representation under differing circumstances and offering different analysis possibilities. We shall introduce the basic classes of Petri nets: condition/event, place/transition and predicate/transition nets.

We will first devote a section to the definition of *nets* — the common conceptual basis of all Petri net classes.

5.1 Causal Interconnection of State and Transition Elements: Nets

Nets are bipartite digraphs together with a "weak interpretation". The given interpretation is called "weak", because any use of a net will require further — compatible — interpretation.

More precisely, we call a *net* a triple $N = (S, T, F)$ which represents a system so that:

(i)   S is a finite set which elements represent *state elements,* that is, elementary components of system states.

The elements of S will be called *S-elements* and will be represented graphically by circles.

(ii) T is a finite set which elements represent *transition elements*, that is, elementary components of state transitions.

The elements of T will be called *T-elements* and will be represented graphically by boxes.

The elements of $S \cup T$ will be called the *elements of the net.*

(iii) F is a finite set of *ordered pairs* of two types — *(S-element, T-element)* and *(T-element, S-element)* — representing causal connections. The first element represents a system component which is an immediate cause for the system component represented by the second element.

F is called the *flow relation of the net.*

Pairs above are graphically represented by directed arcs.

Moreover, we assume that:

(iv) $S \cap T = \emptyset$ and $S \cup T \neq \emptyset$;

(v) digraph $(S \cup T, F)$ is connected;

(vi) every net element — S or T — belongs to at least one pair in F.

Figure 5.1 shows the graphical representation of a net.

Net 5.1 represents a system, and its S-elements   — a, b, c, d, e, A, B, C, D, E, DE, ABC and ABCDE — represent the elementary components used for expressing system states. The T-elements of the net — 1, 2, 3, 4, 5, 6, 7, 8 and 9 — represent the elementary components of system state transitions.

The arcs represent causal relationships. For instance, both the S-elements A and B are causal input to the T-element 6; the S-element AB is causal output of 6. T-element 4 is causal input to S-element E, while T-element 7 is causal output of E.

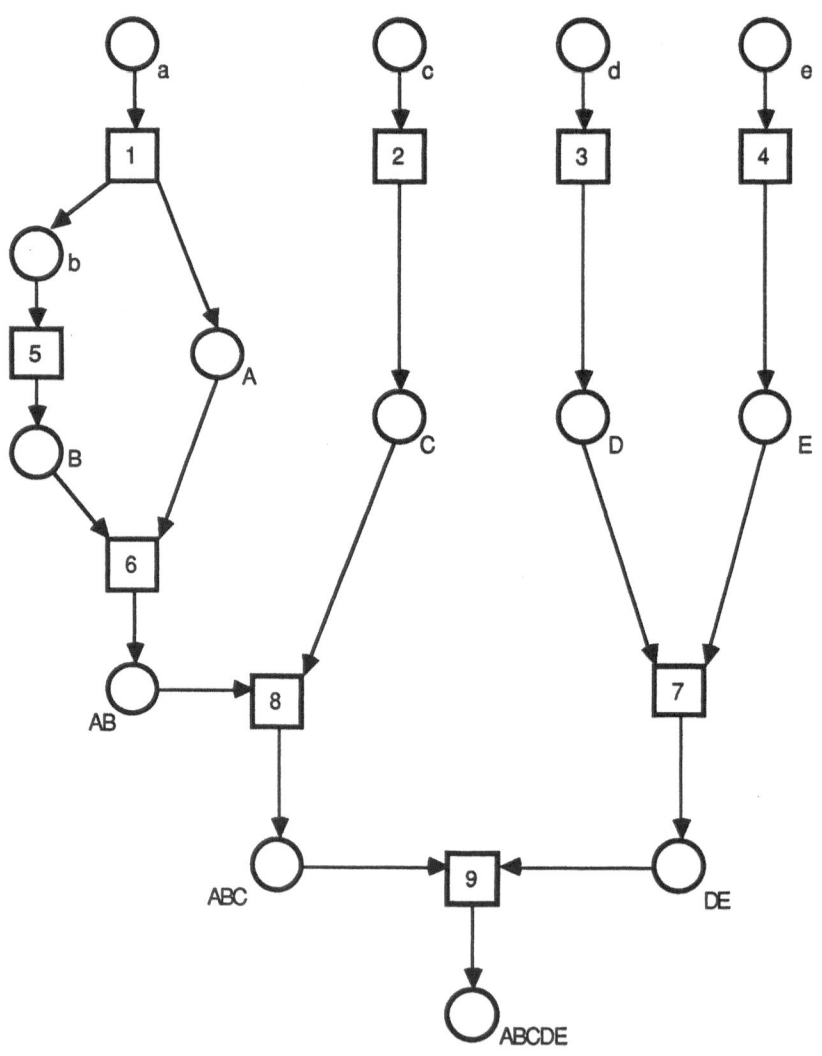

Fig. 5.1

If net 5.1 is to represent an actual system, a more specific interpretation will be necessary. This further net interpretation must harmonize with the "weak interpretation" above.

Therefore, S-element B will always be interpreted by a static component of the actual system: a condition, a resource, a predicate — and never as a dynamic component, such as an event, a transfer, a changing. Conversely, T-element 4 will represent a dynamic component, and not a static one.

Net 5.1 can be viewed as the net representation of a system we already met in Chapter 2, the project plan described by Specification 2.1. Table 5.1 shows this net interpretation, which is perfectly compatible with the "weak interpretation" associated with all nets, as the reader can easily check.

Table 5.1

T-element	Interpretation	S-element	Interpretation
		a	ready to develop A
1	develop A	b	ready to develop B
5	develop B	c	ready to develop C
2	develop C	d	ready to develop D
3	develop D	e	ready to develop E
4	develop E	A	A developed
6	assemble A and B	B	B developed
8	assemble AB and C	C	C developed
7	assemble D and E	D	D developed
9	assemble ABC and DE	E	E developed
		AB	A and B assembled
		DE	D and E assembled
		ABC	AB and C assembled
		ABCDE	ABC and DE assembled

As we see, Figure 5.1 and activity network 2.14 are quite similar. In both cases simple causal precedence dominates the representation. Nevertheless, some interesting differences emerge at once. Many more will emerge later on, when we become familiar with all features of Petri nets.

This view of event 4 partially hides the causal structure of the plan. Indeed, by Specification 2.1, the assembly of A and B may terminate prior to the development of C and enter a waiting state, or vice versa. Such waiting states cannot be represented in activity networks. Analogously, the development of A supplies causal input — the resource "module A" — to the assembly of A and B, input which again cannot be modeled in activity networks.

In nets, waiting states and available resources are represented by means of S-elements. In net 5.1, the S-elements AB, C and A respectively represent the waiting of assembly AB, of assembly C, and of resource A.

In Petri nets, ready-to-go and termination states are made explicit; in activity networks they are not. In net 5.1 the initial state is represented by means of four S-elements — a, c, d and e — one for each ready-to-go activity. This allows us to model the circumstance that some ready-to-go activities may actually not start. In this way, the plan more truly reflects the actual execution possibilities.

Enabling, Steps, Reachability

Let $N = (S, T, F)$ be a net and x an element of N. We call *pre-set* of x the set
$$\bullet x := \{y \mid (y, x) \in F\};$$
we call *post-set* of x the set
$$x \bullet := \{y \mid (x, y) \in F\}.$$
In net 5.1: $\bullet ABC = \{8\}$, $ABC \bullet = \{9\}$, $\bullet 6 = \{A, B\}$, $6 \bullet = \{AB\}$, $\bullet 0 = \emptyset$.

The pre-set of x is the set of all net elements which are input to x, the post-set of x is the set of all net elements which are output of x.

Both the notions of pre-set and of post-set are extended to sets of net elements. The pre-set of the set X of net elements is defined as
$$\bullet X := \{y \mid x \in X \wedge (y, x) \in F\}.$$
Analogously for the post-set of a set X of net elements.

An S-element s is called a *conflict* of N if $\quad \|\bullet s\| \geq 2 \quad \vee \quad \|s \bullet\| \geq 2.$

Now let C be a non-empty subset of S, and t a T-element. We say that t is *enabled at* C — or, *has concession in* C— if $C \supseteq {}^\bullet t$ and $C \cap t^\bullet = \emptyset$.

More simply, t has concession at C if all its inputs belong to the pre-set of t while all its outputs do not. The modeling power of the notion of concession will become apparent in conjunction with using subsets C to represent system states.

Two T-elements t' and t" are *concurrently enabled at* C if they are both enabled at C and do not share input or output S-elements:

$$({}^\bullet t' \cup t'^\bullet) \cap ({}^\bullet t" \cup t"^\bullet) = \emptyset.$$

If t' and t" are both enabled at C but

$${}^\bullet t' \cap {}^\bullet t" \neq \emptyset \quad \vee \quad t'^\bullet \cap t"^\bullet \neq \emptyset$$

we say that t' and t" are *in conflict at* C.

Let C' and C" be two non-empty subsets of S, and E a non-empty subset of transitions enabled at C. We say that C" *is reachable from* C' *by step* E if:

(i)   each pair of T-elements t' and t" belonging to E are concurrently enabled at C';

(ii) $C" = (C' \setminus {}^\bullet E) \cup E^\bullet$.

When C" is reachable from C' by step E we write C'[E>C".

Notice that the former definition is the definition of a binary relation r in $\mathbb{P}(S)$, the power set of S:

$$\forall\ C', C" \in \mathbb{P}(S): \quad (C', C") \in r \ \leftrightarrow\ \exists E \in \mathbb{P}(T): C'[E>C".$$

r is called the *one-step reachability relation* of the net.

The relation $r^* := (r \cup r^{-1})^*$ — where the star indicates finite iteration — is called the *reachability relation* of the net. If $(C', C") \in r^*$ we say that C" is *reachable from* C'.

A pair (C', C") of subsets of S belongs to r^* if and only if C" can be reached from C' in a finite number of forward or backward steps.

r^* is an equivalence relation. Indeed, it is quite easy to see that relation r^* is:

reflexive: $\forall\ C \in \mathbb{P}(S): \quad (C, C) \in r^*$,

symmetric: $\forall\ C', C" \in \mathbb{P}(S): \quad (C', C") \in r^* \rightarrow (C", C') \in r^*$,   and

transitive: $\forall\ C', C", C''' \in \mathbb{P}(S): \quad (C', C"), (C", C''') \in r^* \rightarrow (C', C''') \in r^*$

Pure Nets, Simple Nets

We say that an S-element s is a *side-condition* of a T-element t if
$$s \in \cdot t \cap t \cdot.$$

A net without side-conditions is said to be *pure*. Net 5.1 is pure. In Fig. 5.2, the S-element A is a side-condition of the T-element 1.

We say that N is *S-simple* if
$$\forall\, s', s'' \in S: \quad \cdot s' = \cdot s'' \ \wedge \ s' \cdot = s'' \cdot \quad \rightarrow \quad s' = s'',$$
and that N is *T-simple* if
$$\forall\, t', t'' \in T: \quad \cdot t' = \cdot t'' \ \wedge \ t' \cdot = t'' \cdot \quad \rightarrow \quad t' = t''.$$

N is called *simple* if it is both S-simple and T-simple. In a simple net an element is completely identified by its input and output: no two distinct elements have the same input and output. Nets 5.1 and 5.2 are simple.

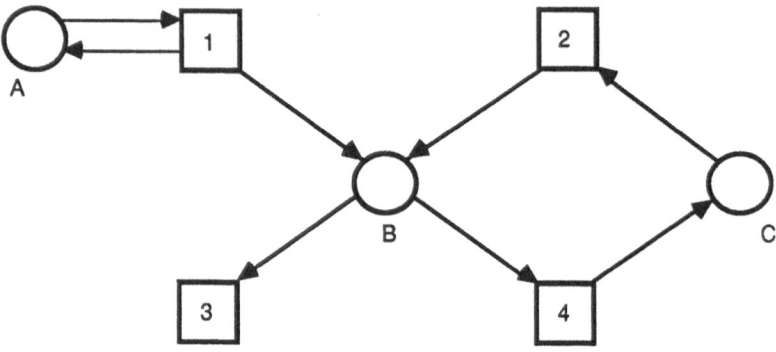

Fig. 5.2

S-Graph, T-Graph, Free Choice Net

A net $N = (S, T, F)$ is called an *S-graph* or a *state machine* if
$$\forall\, t \in T: \quad /\!/ \cdot t/\!/ = /\!/ t \cdot /\!/ = 1,$$
and is called a *T-graph* or a *marked graph* if
$$\forall\, s \in S: \quad /\!/ \cdot s/\!/ = /\!/ s \cdot /\!/ = 1.$$

One can eliminate all the T-elements of an S-graph graphically without loss of meaning by replacing each instance of

The result is a digraph whose vertices correspond to the S-elements of the net. If we operate similarly on the S-elements of a T-graph, we get a digraph whose vertices correspond to the T-elements of the net.

Figures 5.3 and 5.4 respectively show an S-graph and a T-graph. Figure 5.2 shows a net which is neither an S-graph nor a T-graph.

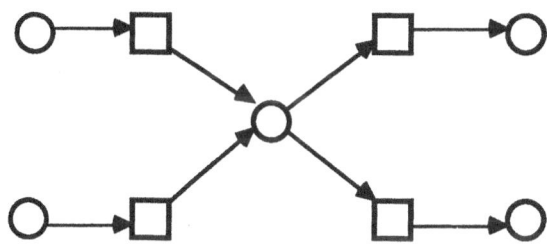

Fig. 5.3

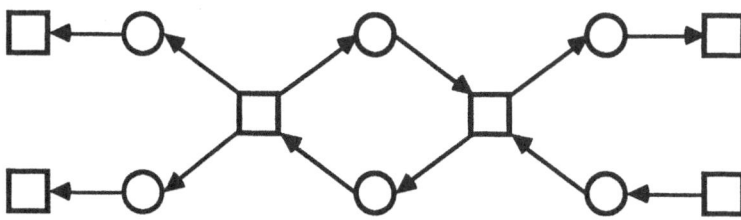

Fig. 5.4

We say that N is a *free choice net* if
$$\forall\, s \in S : \quad \| s \cdot \| > 1 \;\rightarrow\; \cdot(s\cdot) = \{s\}.$$

All nets seen up to now are free choice; the net in Fig. 5.5 is not.

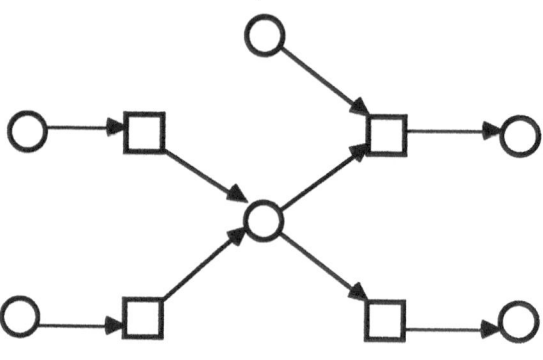

Fig. 5.5

We will come back to S-graphs, T-graphs and free choice nets in the sequel, when discussing the analysis possibilities of Petri net models.

Subnets, S-Vectors, T-Vectors

We say that net $N' = (S', T', F')$ is a *subnet* of net $N = (S, T, F)$ if
$$S \supseteq S', \quad T \supseteq T', \quad F' = F \cap (S' \times T' \cup T' \times S').$$

N' is a subnet of N if its elements belong to N and its flow relation is the restriction of the flow relation of N to those common elements.

The subnets of net N are bipartite sub-digraphs induced in N by any subset of vertices. A subnet comprehends all arcs joining its nodes in the embedding net. Net 5.3 is a subnet of net 5.4.

Now assume a strict ordering has been fixed for both the place and transition sets of net $N = (S, T, F)$:
$$S : \quad s_1 < s_2 < \dots < s_m,$$
$$T : \quad t_1 < t_2 < \dots < t_n.$$

An *S-vector* is any row or column vector indexed by means of the elements of S, and a *T-vector* is such a vector indexed by T.

Sets of places can be represented by means of S-vectors of zeros and ones. In net 5.1, for instance, if we assume the ordering   a < A < AB < ABC < ABCDE < b < B < c < C < d < D < DE < e < E, set {a, c, d, e} may be represented by means of S-vector [1, 0, 0, 0, 0, 0, 0, 1, 0, 1, 0, 0, 1, 0] .

Markings are usually represented as S-vectors whose entries are the markings of the corresponding places.

We say that the subnet $N' = (S', T', F')$ of net $N = (S, T, F)$ is an *S-subnet* if

(i) $T' = \cdot S' \cup S' \cdot$

(ii) $\forall t \in T': \; /\!/ \cdot t \cap T' /\!/ = /\!/ t \cdot \cap T' /\!/$

where the pre- and post-set operators refer to N.

S-subnets are subnets whose T-elements are all and only the original inputs and outputs of their S-elements, and which have as many inputs as outputs. S-components are identified by the set of their S-elements, and can be represented by means of S-vectors of zeros and ones.

We say that subnet $N' = (S', T', F')$ of net $N = (S, T, F)$ is a *T-subnet* of N if

(i) $S' = \cdot T' \cup T' \cdot$

(ii) $\forall s \in S': \; /\!/ \cdot s \cap S' /\!/ = /\!/ s \cdot \cap S' /\!/$

again with all pre- and post-set operators referring to N.

T-subnets are subnets whose S-elements are all and only the original inputs and outputs of their T-elements, and which have as many inputs as outputs. They are identified by the set of their T-elements, and may be represented as T-vectors of zeros and ones.

We will meet S- and T-subnets again further on , when discussing the S- and T-invariants of place/transition nets.

5.2 Causal Interconnection of Conditions and Events: CE Nets

Condition/event nets — CE nets — build the fundamental class of Petri nets. They offer a unitary starting point for understanding the other Petri net models and their relationships.

Condition/event nets offer a suitable plan representation tool whenever we are interested in clock-independent analysis of plan states and their transitions. CE nets are nets, and therefore made up of S- and T-elements. But their definition explicitly refers to global system states and global state changes.

As we will soon see, the definition of CE nets involves a special interpretation, in the same sense that nets as introduced above entail an interpretation. The interpretation associated with CE nets conforms to the interpretation associated with nets in general.

A *condition/event net* — a *CE net* — is a quadruple $Q = (S, T, F, C)$ which satisfies the following hypotheses:

(i) (S, T, F) is a simple net;

(iii) The S-elements of (S, T, F) represent elementary system conditions, while its T-elements represent elementary system events.

S-elements of CE nets are called *conditions,* T-elements are called *events.*

(ii) C is a non-empty set of subsets of S — called *cases* — such that:
-   C is an equivalence class of the reachability relation $r*$;
-   every T-element is enabled in at least one case.

Cases represent maximal sets of together holding conditions.

C is called the *case class* of Q.

A CE net $Q = (S, T, F, C)$ is graphically represented like net (S, T, F) but for the representation of cases. Being an equivalence class C is fully identified by one of its elements — that is, by any case.

Cases are sets of conditions. We graphically represent cases drawing a dot into the circles representing their conditions. Such dots are also called *tokens.* It is customary to view the graphically represented case as *the actual case.*

Figure 5.6 shows the graphical representation of a CE net $Q = (S, T, F, C)$ where:
-   $S = \{A, B, C\}$ and $T = \{1, 2, 3, 4\}$
-   $F = \{(A, 1), (1, B), (B, 3), (3, A), (C, 2), (2, B), (B, 4), (4, C)\}$
-   $C = \{\{A\}, \{B\}, \{C\}\}$
-   the actual case is $\{B\}$.

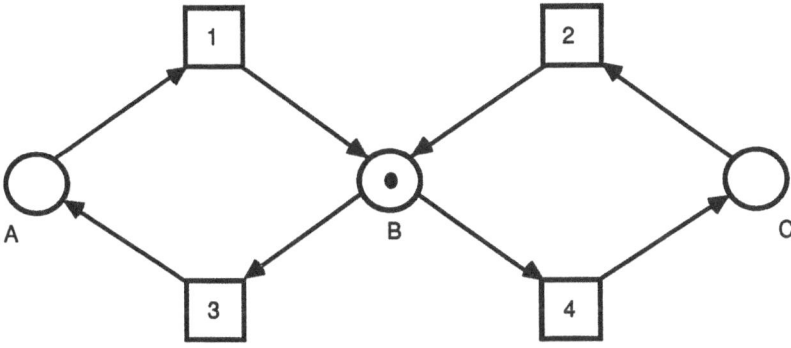

Fig. 5.6

The definition of condition/event nets alone does not lend itself to useful applications: a further piece of interpretation is necessary.

At this point, we have conditions, events, flow relation and a case class; we also introduced the definition of step, of one-step-reachability and of reachability as relation among cases. But in order to work with CE nets, we need the following understandings:

- We take cases to represent system states; the S-elements belonging to a case represent conditions which hold true in the system state represented by the case. The S-elements are therefore called conditions.

- The actual state of the system is represented by a variable over the case class C. Let K be a variable over C representing the actual system state.

- Let C be a case, and set the variable K:= C. At case K, enabled events may occur. An event e is enabled at case K if all conditions $c \in \cdot e$ belong to K, while all conditions $c \in e\cdot$ do not. After the occurrence of e, conditions $c \in \cdot e$ no longer belong to K, while conditions $c \in e\cdot$ do. The value of K will be changed:

$$K = (C \backslash \cdot e) \cup e\cdot = C' \in C.$$

C' is *a follower case of* case C.

- It is not specified if an enabled event will ever occur, nor when.

- Steps are defined as sets of events which are concurrently enabled at a given case. A step represents a non-elementary state transition whose elementary components may take place in a causally independent way. Observe that steps are *not* defined to be maximal sets of concurrently enabled events.

- Steps *may be executed* or not. The execution of a step changes the actual case K.

- Since C is an equivalence class of relation **r***, every case is reachable from any other case after a finite number of forward or backward steps. Not every case, of course, is reachable from any other case by forward steps. C represents a maximal set of system states, one reachable from another by a finite number of forward or backward state transitions.
- The equivalence classes of relation **r*** different from C represent sets of system states which cannot be reached starting with cases belonging to C.

In the CE net 5.6, the actual case is {B}. Condition B holds, conditions A and C do not. Events 3 and 4 are enabled, events 1 and 2 are not. If event 3 occurs the actual case will be {A}, and event 1 will be enabled. If event 4 does, the actual case will be {C}, and event 2 will be enabled. At case {B}, there are two steps — {3} and {4} — both consisting of a single event.

Figure 5.7 shows a condition/event net based on net 5.1. The illustrated case {b, A, C, DE} is to be interpreted as: modules A and C have been developed, development of module B is ready to start, modules D and E have been assembled.

The case class of CE net 5.7 is:

{ { a, c, d, e }, { b, A, c, d, e }, { a, C, d, e }, { a, c, D, e }, { a, c, d, E },
 { b, A, C, d, e }, { b, A, c, D, e }, { b, A, c, d, E }, { b, A, C, d, E },
 { b, A, C, D, e }, { b, A, c, D, E }, { b, A, C, D, E }, { b, A, c, DE }, { b, A, C, DE },
 { a, C, D, e }, { a, C, d, E }, { a, c, D, E }, { a, C, D, E }, { a, c, DE }, { a, C, DE },
 { B, A, C, d, e }, { B, A, c, D, e }, { B, A, c, d, E }, { B, A, C, d, E }, { B, A, C, D, e },
 { B, A, c, D, E }, { B, A, C, D, E }, { B, A, c, DE }, { B, A, C, DE },
 { AB, C, d, e }, { AB, c, D, e }, { AB, c, d, E }, { AB, C, d, E }, { AB, C, D, e },
 { AB, c, D, E }, { AB, C, D, E }, { AB, c, DE }, { AB, C, DE }, { ABC, d, e },
 { ABC, d, E }, { ABC, D, e }, { ABC, D, E }, { ABC, DE }, { ABCDE } } ,

where each case represents an execution state of the project.

At case { b, A, C, DE } in the figure only step {5} is executable; at case { B, A, C, D, E } the executable steps are {6}, {7}, {6, 7}.

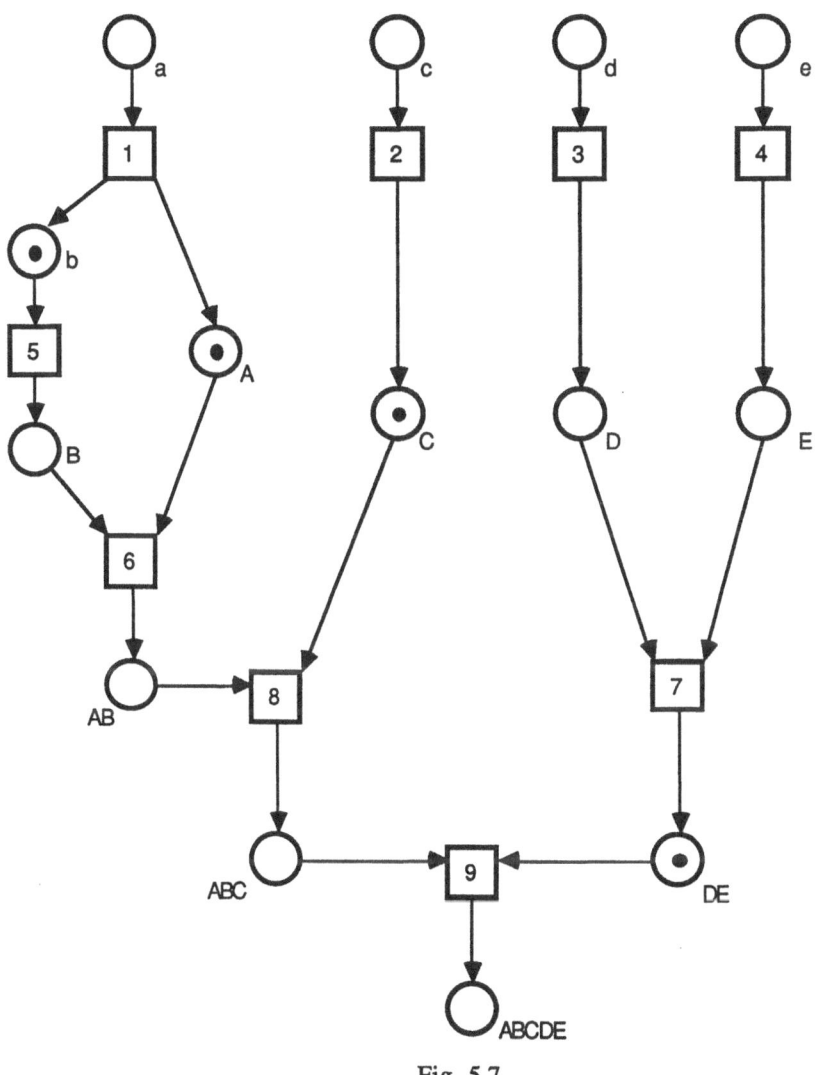

Fig. 5.7

About Petri Net Plans

At the beginning of this chapter we explained that speaking of plan execution states implies the existence of a plan supervisor. No supervisor is however explicitly represented in Petri net plans: both the existence of system states and the taking place of state transitions are simply assumed.

In Petri net plans, elementary state transitions which can be activated at a certain state will not necessarily be. It is not specified when, if ever, enabled state transitions components will be activated, and for what reason. *The taking place of state transitions is not planned.* The causal structure of state transitions — pre and post-conditions — is described, but the control over the occurrence of state transitions is not specified by the plan, rather, it is assumed to lie in the environment in which the plan is executed.

Since the plan environment is postulated to exist but not described, state transitions are represented as the result of local, spontaneous happenings. This makes Petri net plans suitable for modeling distributed systems.

Only the plan environment exerts influence over plan executions. The boundary between plan and plan environment depends on the use to which the plan is put. What is included in the plan and what is left to its environment is a choice the planner makes in how much to pre-specify. Therefore, a more detailed version of the plan may change this balance.

5.3 Causal Interconnection of Resources and Operations: PT Nets

Place/transition nets are the the most widely applied and the best studied class of Petri nets, so that they are often called Petri nets tout court. Even so, place/transition nets may be seen as a shorthand for condition/event nets, in the sense that every place/transition net can easily be unfolded into a condition/event net.

The reason for the broader diffusion of place/transition nets is that their formalism better respects our habits about system description. An initial system state is directly assigned, together with the rule for generating the possible system behaviors. Compared to the homologous mechanism provided by CE nets — case class plus reachability relation — this more procedural approach comes closer to our usual ways of describing system behaviors. For instance, it comes closer to the interpretation of activity networks, or to the way flow charts of computer programs are conceived.

Moreover, place/transition nets make the representation of multiple resources natural, as S-elements represent "places" where multiple copies of resources may be stored.

To the project engineer, place/transition nets offer the means for designing project plans in which the resource flow among activities is neatly represented. The T-elements will be interpreted as elementary operations connected via resources. The oriented arcs represent the consumption/production of resources by operations.

A *place/transition net* — *PT net* — is a sextuple $N = (S, T, F, K, W, M_0)$ such that:

(i) (S, T, F) is a net where:

S-elements represent *states of elementary resources*, and are called *places;*

T-elements represent *elementary operations*, and are called *transitions;*

the flow relation F represents either *resource consumption* by pairs of type (S-element, T-element), or *resource production* by pairs (T-element, S-element).

For simplicity, we often will call the state of a certain elementary resource, represented by place s, just state s.

(S, T, F) is called the *underlying net.*

(ii) $K : S \rightarrow N \cup \{\aleph_0\}$ is a mapping which associates either a natural number or $\aleph_0$, the cardinality of set of naturals, to each place of the net.

K is called *capacity function* of the net. The value K(s) is the *capacity of place* s.

K(s) represents the highest number of resource units allowed to be simultaneously in state s. If $K(s) = \aleph_0$, that number is unlimited.

(iii) $W : F \rightarrow N$ is a mapping which associates a natural number to each arc of N.

W is called the *weight function* of the net; the value W (x, y) is called the *weight of arc* (x, y).

W (x,y) = n is a shorthand for n arcs going from node x to node y.

The weight W (x, y) is inscribed at the arc (x, y), and usually omitted if equal to one.

(iv) Every mapping $M : S \rightarrow \mathfrak{I}^+$ such that $\forall s \in S : 0 \leq M(s) \leq K(s)$ is called a *marking* of the net. M(s) is called a *marking of place* s, and represents a number of resource units actually in state s.

Markings represent actual distributions of resources.

Marking M is graphically represented by drawing M(s) black dots called *tokens* into place s.

M_0 is a special marking called the *initial marking* of the net.

M_0 (s) represents the number of resource units which are initially in status s.

(v)   The following *ransition rule* is defined.

A transition $t \in T$ is said to be *enabled at marking* M, if

$\forall s \in \cdot t : W(s, t) \leq M(s)$ and

$\forall s \in t \cdot : M(s) \leq K(s) - W(t, s)$.

Enabled transitions *may occur*, or *may be activated* .

The occurrence of transition t changes the actual marking M into the marking M' defined by:

$\forall s \in S : M'(s) = M(s) - W (s, t) + W (t, s)$.

We express this state of affairs by writing $M[t > M'$.

Transitions t', ... , t" are said to be *concurrently enabled at marking* M if they are all enabled at M, and after the occurrence of any one of them, the others are still enabled.

Concurrently enabled transitions *may occur concurrently* .

The concurrent occurrence of a multiset of transitions t', ... , t" is called a *step*, and changes the actual marking M into the marking M' defined by:

$\forall s \in S : M'(s) = M(s) - W (s, t') - ... - W (s, t") + W (t', s) + ... + W(t", s)$.

We express this state of affairs by writing: $M[t', ... , t" > M'$.

Observe that t' = t" is allowed: transitions may occur concurrently to themselves.

An enabled transition represents an elementary operation whose necessary resources are all available, and its occurrence represents the execution of that elementary operation. The execution of an operation changes the distribution of resources.

The PT net $N = (S, T, F, K, W, M_0)$ is said to be *pure* if its underlying net is pure.

Given a PT net $N = (S, T, F, K, W, M_0)$, we denote by $[M_0 >$ the smallest set of markings such that:

(i) $M_0 \in [M_0 >$

(ii) $M' \in [M_0 > \wedge \exists t \in T : M'[t > M" \rightarrow M" \in [M_0 >$.

We call $[M_0 >$ the *set of forward reachable markings*. Moreover, we denote by $[M_0]$ the smallest set of markings such that:

(i) $M_0 \in [M_0]$

(ii) $M' \in [M_0] \wedge \exists t \in T : M'[t > M" \rightarrow   M" \in [M_0]$

(iii) $M" \in [M_0] \wedge \exists t \in T : M'[t > M" \rightarrow   M' \in [M_0]$.

We call $[M_0]$ the *set of forward and backward reachable markings*.

Forward reachable markings represent resource distribution states which the system may enter in its future life. Backward reachable markings represent resource distribution states in which the system could have been in its past life. These latter may be investigated in order to understand how a specific state has been reached. This information could help in achieving a desired situation or avoiding an undesired one.

Figure 5.8 shows a place/transition net with its initial marking; Figure 5.9 illustrates the marking reached by activating first transition 1 and then transition 2. The capacity of places A and B is $\aleph_0$. Any capacity greater than one would not change the behavior of this net, but capacity one for place A would not be consistent with the given initial marking, while capacity one for place B would "block" the net at the initial marking.

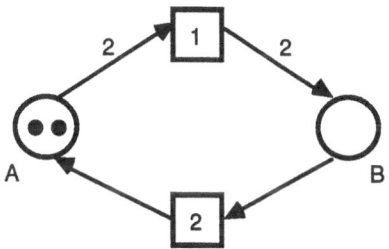
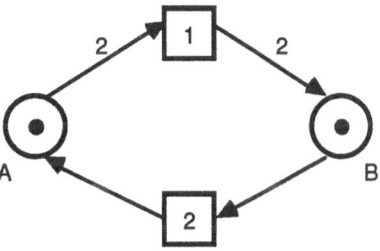

Fig. 5.8   Fig. 5.9

A place p is called a *conflict* if
$$\//\bullet p\// \geq 2 \quad \vee \quad \//p\bullet\// \geq 2.$$

We say that *two transitions* are *in conflict at marking* M if they are both enabled under M but after the occurrence of either transition the other is no longer enabled.

In net 5.6 — now interpreted as a PT net with its initial marking — transitions 3 and 4 are in conflict: if 3 occurs, 4 is no longer enabled; if 4 occurs, 3 is no longer enabled.

A transition t is called a *synchronization transition* if
$$\//\bullet t\// \geq 2 \quad \vee \quad \//t\bullet\// \geq 2.$$

Two places are said to be *synchronized at marking* M if they are two inputs or two outputs of t. In synchronized places resources will be consumed or produced together.

In the net 5.1 — interpreted as a PT net — transition 1 synchronizes places b and A: the ready-to-develop-B signal is "delivered" together with module A. Transition 7 synchronizes places D and E: module D and module E are "consumed" together.

Figure 5.10 shows the unfolding of PT net 5.8 into a condition/event net representing the same plan. The case class of the CE net is { {A', A"}, {B', B"}, {A', B"}, {B', A"} }, and can be seen as an unfolding of set $[M_0]$.

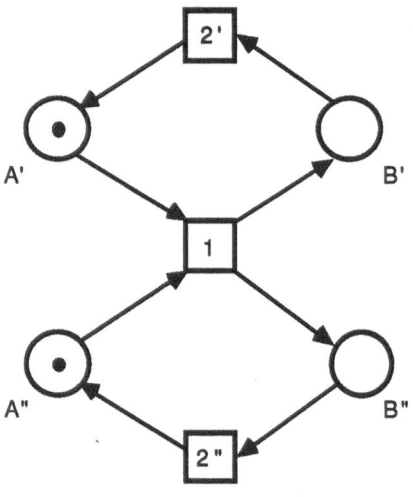

Fig. 5.10

Place/transition nets for which both capacity and weight are constant mappings to the value one, are isomorphic to CE nets. So, for instance, the diagram of Fig. 5.7 can be interpreted both as a CE and as a PT net.

Therefore, one could think that using either net model is indifferent, at least whenever place capacities and arc weights are equal to one. But figuring out a project plan as a structured set of elementary state and state transition components will in general lead to a completely different net than designing the same plan as a set of operations interrelated by means of resources. In fact, plans are descriptions of the work intended by a certain project, work which can usually be conceived in many different ways.

Also, unfolding place/transition nets into condition/event nets representing the same plan is always possible. This is, however, of little practical utility both because the resulting CE net will usually be definitely larger than the PT net, and because the resulting case class will be too large to allow any insight. In applications, and particularly in the design of project plans, initial marking plus transition rule offer a more compact and handy description of the net behavior.

In the next sections we shall introduce the fundamental tools for the structural analysis of place/transition nets. All PT nets in the rest of this chapter are assumed to have unbounded place capacity. They will therefore be denoted by quintuples of the form $N = (S, T, F, W, M_0)$.

We shall use strings of characters — letters and digits — to denote S- and T-elements, and assume that S and T are lexicographically ordered with the succession of natural numbers appended at the tail of the alphabet. S-vectors representing markings will be printed in bold-face.

Incidence Matrix

The structure of a pure place/transition net can be algebraically represented by means of the so-called incidence matrix.

Let $N = (S, T, F, W, M_0)$ be a PT net whose underlying net is pure with S and T strictly ordered. We call *incidence matrix* of net N the matrix $W = [w_{st}]$ defined by

$$w_{st} := \begin{cases} 0 & \text{if} & s \notin {}^\bullet t \cup t^\bullet \\ -W(s, t) & \text{if} & s \in {}^\bullet t \\ W(t, s) & \text{if} & s \in t^\bullet \end{cases}$$

where s and t are indices over the ordered sets S and T.

All zero entries omitted, the incidence matrix of PT net 5.7 is:

W	1	2	3	4	5	6	7	8	9
a	-1								
A	1					-1			
AB						1		-1	
ABC								1	-1
ABCDE									1
b	1				-1				
B					1	-1			
c		-1							
C		1						-1	
d			-1						
D			1				-1		
DE							1		-1
e				-1					
E				1			-1		

The incidence matrix of PT net 5.8 is:

W	1	2
A	−2	1
B	2	−1

Each column of the incidence matrix corresponds to a transition, and expresses the variation induced in the distribution of tokens over the net by activating that transition once.

For any two markings M and M' such that $M[t >M'$, $M' = M + w_t$ holds true, where w_t is the column of W corresponding to transition t, and M' and M are S-vectors respectively representing markings M' and M.

Each row of the incidence matrix corresponds to a place, and records the resource claims of net transitions to that place — negative when the transition takes, positive when it puts.

Incidence matrices fully characterize the graph structure of pure PT nets: two pure PT nets are isomorphic weighted bipartite digraphs if and only if — but for the ordering of S and of T — they have the same incidence matrix.

S- and T-subnets of special interest in net analysis can be found by means of the incident matrix. They are called net invariants, and we shall discuss them in Sect. 5.5. Before doing so, we will introduce some other conceptual tools for the analysis of place/transition nets.

Reachability Graph

A complete representation of the states and state-transitions of a place/transition net is supplied by a graph with vertices labelled by markings and arrows by steps.

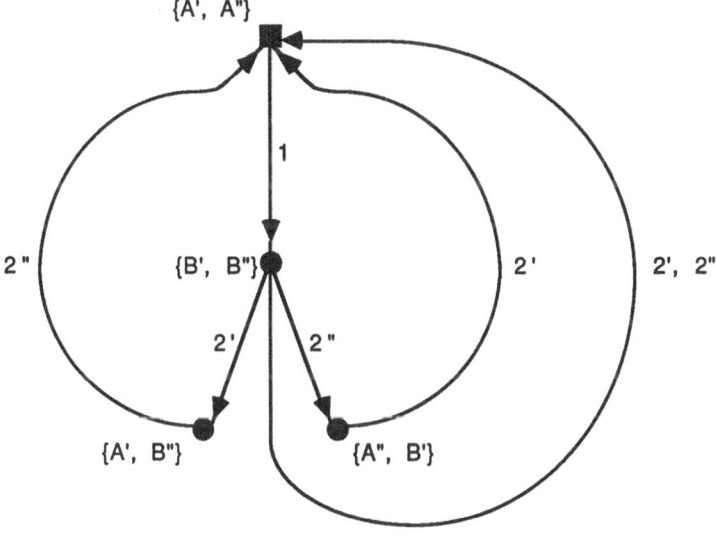

Fig. 5.11

The *reachability graph* of a PT net $N = (S, T, F, W, M_0)$ is a rooted graph defined by the following constructive procedure:

1. Draw the root, and label it M_0 ;
2. if no step is enabled at marking M_0 then end

else: for every step $M_0[t', \dots, t''> M$ draw a vertex labelled M and an arc directed from M_0 to M labelled t', ... , t'';

3. if there is no step enabled at anyone of the markings M above then end
    else: for every step M[t', ... , t''> M' :
    if our graph already contains a vertex labelled M' then draw an arc from M to M' labelled t', ... , t''
    else draw a vertex labelled M' and an arc from M to M' labelled t', ... , t'';

4. repeat from 3.

Recall that by our definition of step transitions t', ... , t'' do not need to be distinct. Figure 5.11 shows the reachability graph of PT net 5.10, Figure 5.12 that of PT net 5.7. The first one is *covered by cycles* — i. e., every vertex lies on a cycle — and every cycle contains the root. This means that net 5.10 has no end marking, and that reachable markings can reached again and again. The represented system has no end state, and possible system states can be achieved from every other possible state.

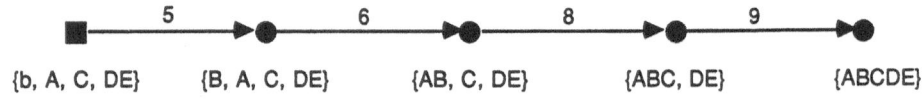

Fig. 5.12

The reachability graph of Fig. 5.12 is a finite *path of length* 4 beginning with the root, { b, A, C, DE}, and ending with vertex { ABCDE }. In net 5.7 there is both an initial and an end marking, and no two transitions are concurrently enabled or in conflict. In other words, the represented system has both an initial and a final state, and encompasses neither resource independent operations nor conflicts.

Algorithms for constructing reachability graphs are complex. It has been shown [Li] that for place/transition nets reachability is at best decidable within exponential space. Also, it has been proved [MM] that there exists a sequence $N_1, N_2, \dots$ of place/transition nets with linearly growing size — number of elements, arcs and initial tokens — such that the order of corresponding reachability graphs $R_1, R_2, \dots$ grows quicker than any primitive recursive function.

Nonetheless, reachability graphs are the most often applied tool for analyzing place/transition nets. Besides answering reachability questions, they permit one to distinguish between conflicting and concurrently enabled transitions: both of these label arcs going out of one vertex, but conflicting transitions do not label successive arcs whereas concurrently enabled transitions always do.

Reachability graphs are also used for investigating other relevant net features, such as liveness, and presence of stop or home states.

Liveness, Stop States and Home States

Observing the reachability graph of Fig. 5.11 at any given marking, we see that every net transition may be enabled again, at a subsequent marking. This feature is not shared by net 5.7, as the reachability graph of Fig. 5.12 shows.

Given a place/transition net $N = (S, T, F, W, M_0)$, we say that *transition* t *is* *live* if
$$\forall M \in [M_0> \ \exists M' \in [M> : \quad t \text{ is enabled at marking } M';$$
we say that *net* N *is live* if every transition $t \in T$ is live.

A *marking* M of N is called *live* if the PT net (S, T, F, W, M) is live.

Nets 5.8, 5.9 and 5.10 are live, net 5.7 is not. Observe that net 5.7 would be live with a different initial marking — with $M_0 = \emptyset$.

Liveness is a relevant feature: live nets represent plans in which no operation can once and for all drop out. The liveness of general place/transition nets can be established only by constructing the reachability graph. For some special nets — such as T-nets — other, less cumbersome, methods exist. We shall not treat this rather specialist topic here, but suggest reading the related bibliography [La, BS, Lie].

A reachable marking M is called a *stop state* if
$$\forall t \in T : \quad t \text{ is not enabled at marking } M;$$
it is called a *home state* if
$$\forall M' \in [M_0> : \quad M \in [M'>.$$

Stop states represent plan terminations. Marking {ABCDE} of PT net 5.7 is a stop state, and represents the only planned termination.

Live place/transition nets — like nets 5.8, 5.9 and 5.10 — have no stop states.

Place/transition nets without stop states are not necessarily live. Figure 5.13 shows a non-live PT net without stop states, and its reachability graph. The initial marking is the illustrated one.

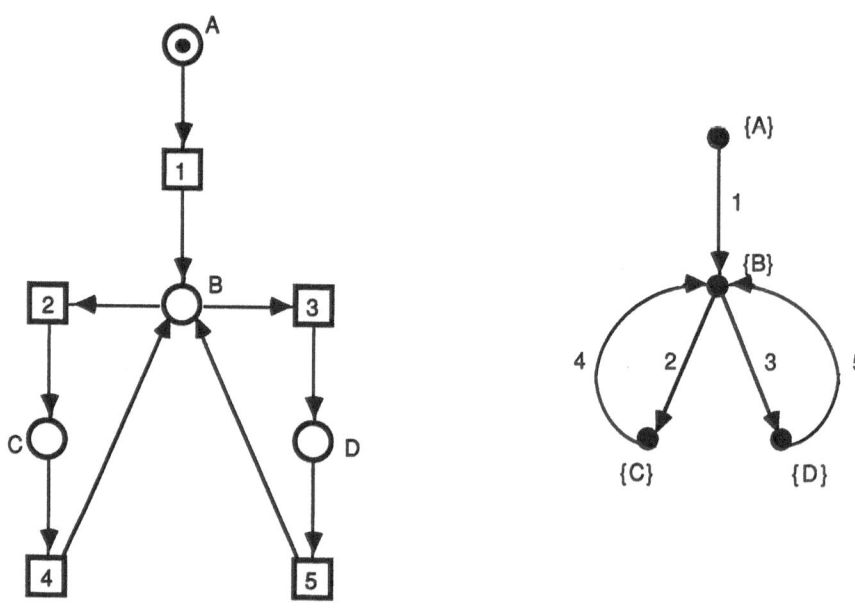

Fig. 5.13

Marking {A, A'} of net 5.10 and marking {B} of net 5.13 are home states. As we see, PT nets having home states are not necessarily live.

Place/transition nets with stop states have no home states. Indeed, PT net 5.7 has no home states.

A reachable marking M of the above PT net N is said to be *reproducible* if
$$\exists \; M' \in [M> : \quad M' \neq M \; \wedge \; M \in [M'>.$$

Home states are reproducible markings, but not all reproducible markings are home states. In a project plan represented by means of a place/transition net, home states represent plan execution states which can be repeatedly achieved. They will be a central concern of the next section.

5.4 Conservation of Resource Counts and Reproduction of States : S- and T-Invariants

Consider the place/transition nets of Fig. 5.14. N_1 has the property that whatever goes on in the net, the total token count over places A and B will be equal to two. N_2 does not have this property: any token count is possible for it.

On the other hand, any marking of N_2 can be restored after having been changed, while no restoring is possible for the markings of N_1. Therefore, in N_2 every enabled activating sequence will again be enabled.

PT net N_1 PT net N_2

Fig. 5.14

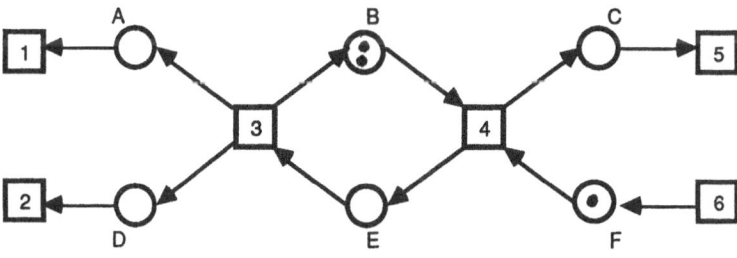

Fig. 5.15

Now observe place/transition net 5.15. The S-subnet identified by places B and E preserves its token count: in every forward or and backward reachable marking of net 5.15, the sum of tokens in places B and E equals two.

In net 5.15, no proper subnet is such that any given net marking can be reproduced by suitably activating its transitions. The net as a whole, however, has this property. For instance, the illustrated marking may be restored by activating transitions 5, 6, 3, 1, 2, after having been changed by activating transition 4.

Places B and E identify an S-invariant of the net, while transitions 4, 5, 6, 3, 1, 2 identify a T-invariant.

We will introduce S- and T-invariants as sets of net elements characterized by some marking independent behavioral features. Later on, we will prove that the S- and T-invariants of a PT net can be computed by solving a homogeneous linear equation system based on the incidence matrix.

S-subnets of place/transition nets S-subnets are defined as for nets, but for the arc weights, the requirement that transitions have as many inputs as outputs must here be understood as taking arc multiplicity into account, and analogously for the T-subnets of place/transition nets.

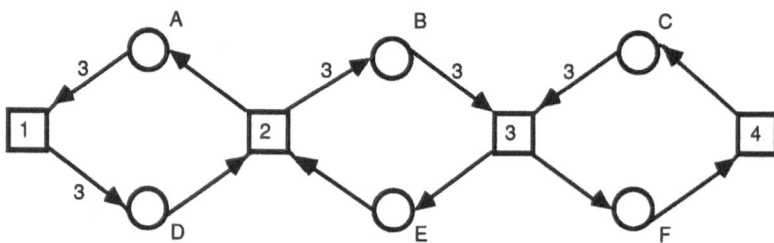

Fig. 5.16

Consider, for instance, the PT net in Fig. 5.16. It has one elementary S-subnet identified by the set of places {A, D}. Its S-vector form is: [1, 0, 0, 1, 0, 0]. The subnet identified by the set of places [0, 1, 0, 0, 1, 0] is not an S-subnet. In that subnet transition 2 has indeed one input arc and three output arcs.

S-subnets are subnets which can neither lose nor produce tokens: by definition, each transition puts as many tokens into the subnet as it takes away. This holds true also for backward activations.

S-weights are S-vectors of integers with entries representing weights for token counts over a given set of places: the entries indicate — for each place — how many times the tokens in that place must be counted. Zero-entries correspond to places not be-longing to the considered set. Negative entries indicate that the token count for the corresponding place must be subtracted. In fact, S-weight are S-vectors of integers to which a particular interpretation has been attached.

In general, the number of tokens in a set of places gets modified when transitions occur. There may, however, be S-weights which yield a constant token count over a given set of places for every reachable marking.

Consider the PT net 5.16 together with any initial marking. [1, 0, 0, 1, 0, 0] is an S-subnet, and therefore the number of tokens in its places — A and D — is constant for any forward or backward reachable marking. The same does not hold true for the number of tokens over the set of places [0, 1, 0, 0, 1, 0]; but the S-weight [0, 1, 0, 0, 3, 0] does yield a constant token count over B and E for any forward or backward reachable marking. On the other hand, the number of tokens in the set of places [0, 0, 1, 0, 0, 1] changes at any forward or backward reachable marking, and no S-weight can keep it constant.

We will call *S-invariant* each S-weight which gives a constant token count for all forward and backward reachable markings. The associated set of places is called the *support* of the S-invariant, and is represented by a vector of zeros and ones.

The non-negative S-invariants whose supports are not properly contained in the support of any other S-invariant are *the elementary S-invariants of the net.*

Net 5.13 has two elementary S-invariants: [1, 0, 0, 1, 0, 0] and [0, 1, 0, 0, 3, 0]. [1, 1, 0, 1, 3, 0] is a non-elementary S-invariant, the "sum" of the two previous ones.

As S-invariants characterize sets of S-elements with invariant weighted token count in all reachable markings, so T-invariants characterize sets of T-elements with invariant occurrence count in all realizable marking reproductions.

T-subnets are subnets which allow for the reproduction of any net marking provided all their transitions are enabled. Marking reproductions are then achieved by activating each transition of the T-subnet once. Indeed, if a transition puts n tokens into a place, there will be another — enabled — transition which can take them away; and vice versa.

T-counts are T-vectors of integers with entries which indicate — for each net transi-tion — how many times it is to be activated in a given occurrence sequence. Zero-entries correspond to transitions not belonging to the considered occurrence sequence. Negative entries indicate backward transition occurrences. Again, T-counts are T-vectors of inte-gers with a peculiar interpretation.

We say that a T-count **j** is *realizable at marking* M if there exists an occurrence sequence whose T-count is **j**, and whose transitions are enabled — one after the other — starting from marking M. Such an occurrence sequence is called a *realization* of **j**.

In general, realizable T-counts will not reproduce net markings. There are, however, T-counts whose realization necessarily leads to marking reproduction.

Consider the PT net in Fig. 5.17. The T-vector [0, 1, 0, 0, 1, 0] identifies a T-subnet: if transitions 2 and 4 are enabled, activating both of them once will give marking reproduction. The T-vector [1, 0, 0, 1, 0, 0] does not identify a T-subnet: activating transitions 1 and 4 once will give no marking reproduction. However, if T-count [1, 0, 0, 3, 0, 0] is realizable, corresponding occurrence sequences will always give a marking reproduction. On the other hand, no T-count over transitions 3 and 4 will ever reproduce a marking.

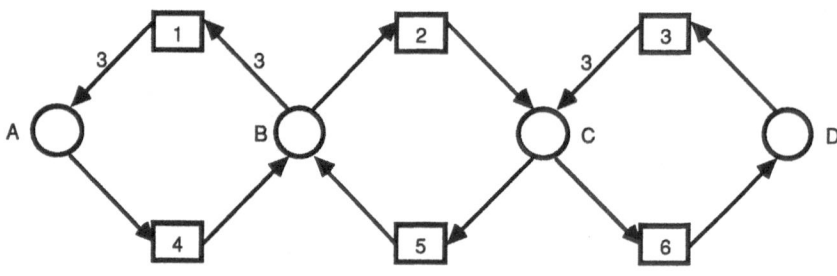

Fig. 5.17

We will call *T-invariants* all T-counts whose realization leads to marking reproduction. The set of transitions belonging to such realizations is called the *support* of the T-invariant, and is represented by a vector of zeros and ones.

The non-negative T-invariants whose supports are not properly contained in the support of any other T-invariant are *the elementary T-invariants of the net*.

PT net 5.17 has two elementary T-invariants: [0, 1, 0, 0, 1, 0] and [1, 0, 0, 3, 0, 0]. [1, 1, 0, 3, 1, 0] is a non-elementary T-invariant, the "sum" of the two elementary T-invariants.

PT net 5.15 has one T-invariant, identified by the T-vector [1, 1, 1, 1, 1, 1]. Occurrence sequences [4, 5, 6, 3, 1, 2] and [4, 6, 5, 3, 2, 1], both of which reproduce the illustrated initial marking, are realizations of this invariant.

PT net 5.7 has neither S- nor T-invariants: no S-weight remains constant at transition occurrences; changed markings can never be restored.

PT net 5.6 has one S-invariant, the S-subnet [1, 1, 1] coinciding with the whole net. The token count over it equals one at any forward or backward reachable marking. This net also has two elementary T-invariants, the T-subnets [1, 0, 1, 0] and [0, 1, 0, 1]. Since they both cover the net and are realizable at any reachable marking, all reachable markings can be restored.

S-invariants identify sets of places over which the weighted count of tokens always remains constant. They represent system parts which are steady with regard to the amount of involved resources, system parts which will neither "lose" nor "gain" resources.

T-invariants identify sets of transitions with multiplicities which are capable of reproducing a starting marking, provided that — from that starting marking — an occurrence sequence with the indicated multiplicities can actually be realized.

Two transitions belonging to different S-invariants are never in conflict for a resource. Therefore, S-invariants represent resource-independent system parts.

Two places belonging to different T-invariants are never both input or output of the same transition. Production/consumption of resources in places of different T-invariants is never synchronized. T-invariants represent operationally independent system parts.

Two transitions of different T-invariants may be in conflict under a certain marking. Such situations represent resource conflicts between operations.

Two places of different S-invariants may be synchronized. Such situations represent synchronization of production/consumption of resources.

So far, we have introduced S- and T-invariants as vectors of integers which express marking-independent behavioral features of PT nets. We have allowed also for S- and T-invariants made up only of zeros, even though they do not have a meaningful interpretation. This will make the formulations in the next sections more elegant.

Calculation of Invariants

In this section we will demonstrate that the S- and T-invariants of a place/transition net can be obtained by linear algebraic calculations based on the incidence matrix. We will assume $N = (S, T, F, W, M_0)$ to be a pure PT net.

We first restate the definitions of S- and T-invariant more formally than above. In the rest of this section, vectors are column vectors and matrix transposition is denoted by the superscript T.

Definition 5.1   We call an *S-invariant* of place/transition net N each S-vector of integers i such that for all $M \in [M_0]$:

$$i^T M_0 = i^T M.$$

Definition 5.2   We call a *T-invariant* of place/transition net N each T-vector of integers j such that its realization as a T-count yields marking reproduction.

The following theorem offers the means for getting the S-invariants of place/transition nets by means of linear algebraic calculations. Like the other theorems in this section, it requires nets to be pure, so that the incidence matrix is defined.

Theorem 5.1   Let $N = (S, T, F, W, M_0)$ be a pure PT net, $W = [w_{st}]$ its incidence matrix, and i an S-vector of integers such that
(*) $i^T W = 0^T$.
Then i is an S-invariant of N.

Proof

If i is an S-vector of integers for which $i^T W = 0^T$, then for each column w_t of W, $i^T w_t = 0$ will hold true.

Now suppose that $M \in [M_0]$. This means that there exists a finite occurrence sequence $(t_1, t_2, \ldots, t_h)$ which changes M_0 into M. The transition occurrences in sequence $(t_1, t_2, \ldots, t_h)$ may be forward or backward.

For any transition t such that $M[t > M'$, $M' = M + w_t$ holds true, where w_t is the column corresponding to transition t in the incidence matrix W. We get:

$$M = M_0 \pm w_{t1} \pm w_{t2} \pm \ldots\ldots \pm w_{th}$$

where minus signs relate to backward activating transitions. Hence:

$$i^T M = i^T M_0 \pm i^T w_{t1} \pm i^T w_{t2} \pm \ldots\ldots \pm i^T w_{th},$$

which gives: $i^T M_0 = i^T M.$

By Def. 5.1, i is an S-invariant of N.   ¶

Observe that Def. 5.1 allows S-invariants i which give $i^T M_0 = i^T M$ for some M $\notin [M_0]$. In fact, this definition just ensures that if for some S-invariant i, $i^T M_0 \neq i^T M$ holds true, then M is not reachable from M_0 — neither forward nor backward.

Theorem 5.2 reverses Theorem 5.1, yet only for special nets.

<u>Theorem 5.2</u>   Let $N = (S, T, F, W, M_0)$ be a pure place/transition net such that
$$\forall t \in T \; \exists M \in [M_0] : \; t \text{ is enabled at marking } M,$$
and let i be an S-invariant of N. Then:
$$i^T W = 0^T,$$
where W is the incidence matrix of N.

<u>Proof</u>
Let $t \in T$, and M be a reachable marking at which t is enabled.

If $M[t > M'$, then $M' = M + w_t$ will hold true, where w_t is the column corresponding to transition t in matrix W.

As i is an S-invariant of N:
$$i^T M_0 = i^T M = i^T M'$$
$$i^T M = i^T (M \pm w_t)$$
(**)
$$i^T w_t = 0^T.$$

Since we required that for every transition $t \in T$ a reachable marking exists at which t is enabled, equality (**) holds true for every column of incidence matrix W. Therefore:
$$i^T W = 0^T. \qquad ¶$$

<u>Theorem 5.3</u>   Let $N = (S, T, F, W, M_0)$ be a pure PT net, and $W = [w_{st}]$ its incidence matrix. A T-vector of integers j such that
(***)
$$W j = 0$$
is a T-invariant of N.

<u>Proof</u>
Let j be a T-vector of integers such that $W j = 0$. We will demonstrate that j is a T-count the realization of which leads to marking reproduction.

For any transition t such that $M[t > M'$, $M' = M + w_t$ holds true, where w_t is the column of W corresponding to transition t.

Now assume that a realization of j starting at marking M exists, and that it yields marking M'. Since j is the count of that realization, we get:

$$M' = M + j_1w_1 + j_2w_2 + \dots + j_hw_h$$

where $w_1, w_2, \dots , w_h$ are the columns of W corresponding to the activated transitions.

Per absurdum, suppose that $M' \neq M$. Then a w_i exists such that:

$$j_1w_{i1} + j_2w_{i2} + \dots + j_hw_{ih} \neq 0.$$

Or equivalently:

$$j^T w_i \neq 0$$

which contradicts our hypotheses.

Hence $M' = M$, which proves that j is a T-invariant of N. ¶

Also Theorem 5.3 can be reversed, but we shall leave the easy proof to the reader. Both the S- and T-invariants of a net can therefore be found by solving a homogeneous linear system based on the incidence matrix. Non-trivial rational solutions can be computed — if they exist — by Gauss elimination. Multiplying them by a common denominator, we obtain integer solutions. But for carrying out net analysis we need the positive S-invariants.

Many algorithms have been proposed for computing generator sets of the elementary S-invariants of place/transition nets [JK]. Unfortunately, the computational burden of all of them is so heavy that even for relatively small nets — nets with 20/30 elements — their computer execution runs into storage and time problems.

All algorithms for computing generator sets of elementary S-invariants have a common theoretical basis. It has recently been proven [CS] that they all are rediscoveries — supplied with ad hoc improvements — of the so-called Fourier-Motzkin method [Fo].

The combinatorial nature of these algorithms is such that performance tests of twelve such algorithms carried out on several nets — each one 20 to 500 elements large — took two moths of CPU time on a VAX 11/750 [CS]. It turned out that one cannot a priori know which one will perform better on a given net: *none* is *best*.

We will show in the next section how the elementary S-invariants can be used for net analysis. But the investment of a non-trivial amount of CPU time must in any case be taken into account.

The calculation of the T-invariants — and, in particular, of a generating system of the elementary T-invariants — is done in a quite similar way, and involves therefore the same problems.

Project Analysis by Means of Invariants

The correspondence between Petri net plans and their specification as to resource flow and iteration of involved actions can be verified by means of elementary S- and T-invariants. We now will demonstrate this technique by means of a small example.

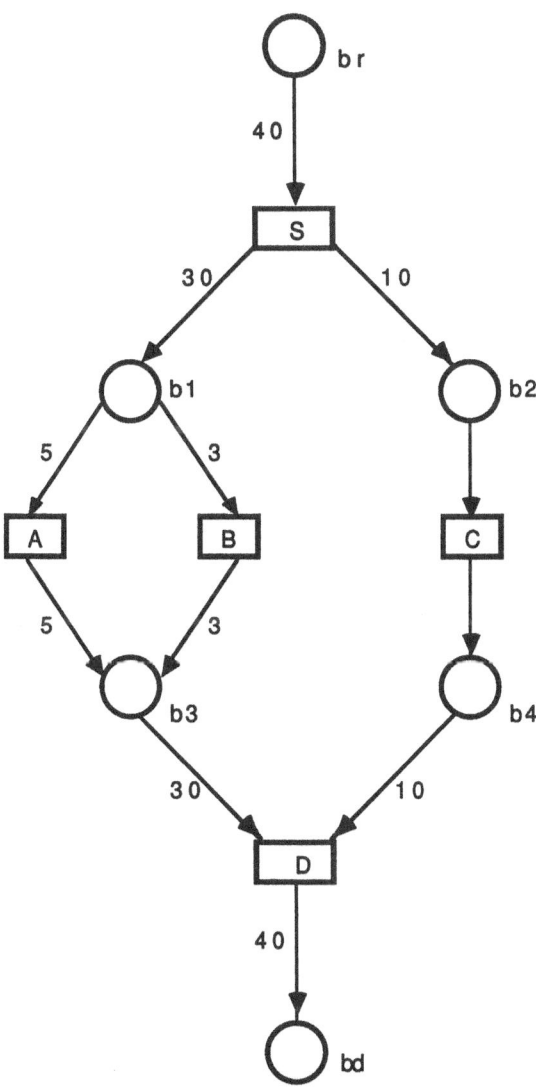

Fig. 5.18

Let the project plan of a car body painting shop be specified as follows:

<u>Specification 5.1:</u>
1. This plan describes the painting of forty car bodies.
2. Forty car bodies will be sent together to the painting shop.
3. Thirty of them will be automatically sprayed on lines A and B; ten will be hand-sprayed, one body at a time.
4. Line A picks up and sprays five car bodies per run; line B three.
5. Once painted, the forty car bodies will be delivered together to the assembly shop.

Figure 5.18 shows a PT net representation of the above specified plan. The interpretation of the net elements is given in Table 5.2. We assume that the initial marking of net 5.18 is: $M_0(\text{br}) = 40$, $M_0(x) = 0$ for all S-element $x \neq \text{br}$.

Table 5.2

Transition	Interpretation	Place	Interpretation
S	send to painting	br	unpainted car bodies
A	spray on line A	b1	bodies for lines A and B
B	spray on line B	b2	bodies to hand-spray
C	hand-spray	b3	bodies for delivery
D	deliver to assembly	b4	bodies for delivery
R	reset	bd	delivered bodies

The incidence matrix of the above net is:

W	S	A	B	C	D	R
br	-40					+40
b1	+30	-5	-3			
b2	+10			-1		
b3		+5	+3		-30	
b4				+1	-10	
bd					+40	-40

where zero-entries have been omitted.

The S-invariants of the net are the integer solutions of the homogeneous linear system

$$[i_1, i_2, \ldots, i_h]\, W = 0^T.$$

The general solution of this system, obtained by Gauss elimination, is:

$$[\,(3h+k)/4\,,\ h\,,\ k\,,\ h\,,\ k\,,\ (3h+k)/4\,]$$

where h and k are rational parameters.

The minimal non-negative integer solutions — and hence the elementary S-invariants of our net — are:

$$i_1 = [1, 0, 4, 0, 4, 1]^T \quad \text{and} \quad i_2 = [3, 4, 0, 4, 0, 3]^T.$$

For any initial marking M_0 and any marking $M \in [M_0]$, by Def. 5.1:

$$i^T M_0 = i^T M.$$

For invariant i_1 and the initial marking represented by the vector M_0 assigned as above, we get:

$$40 = M(br) + 4\, M(b2) + 4\, M(b4) + M(bd)$$

where M is any reachable marking.

Therefore, we can state that for every plan execution starting at M_0:

- if no car bodies are waiting for hand-spraying or delivery, and no car bodies have been delivered to the assembly shop, then forty car bodies are waiting for painting;
- the number of car bodies waiting for hand-spraying or delivery is either zero or ten;
- if no car bodies are waiting for painting, for hand-spraying or for delivery, then forty body have been delivered to the assembly shop.

Similarly, by invariant i_2 :

$$120 = 3\, M(br) + 4\, M(b1) + 4\, M(b3) + 3\, M(bd)$$

which allows us to state that for every plan execution starting at M_0:

- if no car bodies are waiting for automatic spraying or delivery, and no car bodies have been delivered to the assembly shop, then forty car bodies are waiting for painting;
- the number of car bodies waiting for automatic spraying or delivery is either zero or thirty;
- if no car bodies are waiting for painting, for automatic spraying or for delivery, then forty body have been delivered to the assembly shop.

The T-invariants of our net are the integer solutions of the homogeneous linear system

$$W\,[\,j_1, j_2, \ldots, j_h\,]^T = 0,$$

whose general solution — again obtained by Gauss elimination — is:

$$[\,h\,,\ 6h - (3/5)k\,,\ k\,,\ 10h\,,\ h\,,\ h\,]$$

with h and k rational parameters.

Therefore, the minimal non-negative integer solutions of the above system, the elementary T-invariants of our net, are:

$$\mathbf{j_1} = [1, 6, 0, 10, 1, 1]^T, \quad \mathbf{j_2} = [1, 3, 5, 10, 1, 1]^T, \quad \mathbf{j_3} = [1, 0, 10, 10, 1, 1]^T.$$

The elementary occurrence sequences which reproduce the initial marking M_0 represent possible executions of the "whole" plan — that is, complete plan executions.

Therefore, we can state that our net plan has exactly three different complete executions, which all start by sending the car bodies to the painting shop, and all end up with the delivery of the forty painted car bodies to the assembly shop, and the resetting of the plan.

The possible complete executions of our plan are made up of the following actions:

T-invariant $\mathbf{j_1}$:	run line A:	6 times,
	hand-spraying:	10 times;
T-invariant $\mathbf{j_2}$:	run line A:	3 times,
	run line B:	5 times,
	hand-spraying:	10 times;
T-invariant $\mathbf{j_1}$:	run B:	10 times,
	hand-spraying:	10 times.

All plan executions require ten hand-spraying actions. Both executions not involving line B and executions not involving line A are allowed. Executions involving as well line A and as line B are also planned. The plan is flexible with regard to the failure of either of lines A and B, and is not with regard to the failure of hand-spraying activity.

5.5 Representing Operations with Variable Resources Compactly: an Outline of PrT Nets

Place/transition nets tend to become quite large when they represent plans in which a single operation involves different resources on different occasions of its performance. Think, for instance, about modifying the car body painting plan of Specification 5.1 so that "small", "medium" or "large" cars are distinguished. One reason for this could be that operations "spray on line A" and "spray on line B" have a different cost for "small", "medium" or "large" cars.

In the PT net representing our modified plan transitions A and B would have to be substituted by three transitions each, and so would place b1. We see that the size of the new PT net increased considerably.

In a PT net, a place carries one piece of information: the number of available units of the corresponding resource. This information is graphically represented as presence/absence of tokens in the corresponding circle. All other information is expressed by means of the net structure — circles, boxes and arcs. As long as resource units are considered indistinguishable (though countable) entities, the size of a place/transition net only depends on the required detail level. Not so when units of resources come with variable features which make a difference within the framework of the plan. Taking these features into account usually leads to large and complex diagrams — so large and complex as to render both their understanding and manipulation problematic.

Predicate/transition nets (PrT nets) were introduced in order to overcome this difficulty. While place/transition nets tokens are regarded as 'plain' objects, simply being or not being 'there', in PrT nets tokens carry information of various kinds. When building a plan in which it is helpful to think of the production or consumption of resources with variable attributes, PrT nets can be used to great advantage.

Figure 5.19 shows an execution state of the PrT net representing the plan described by Specification 5.2 — a variant of Specification 5.1 where features of resources play a relevant role.

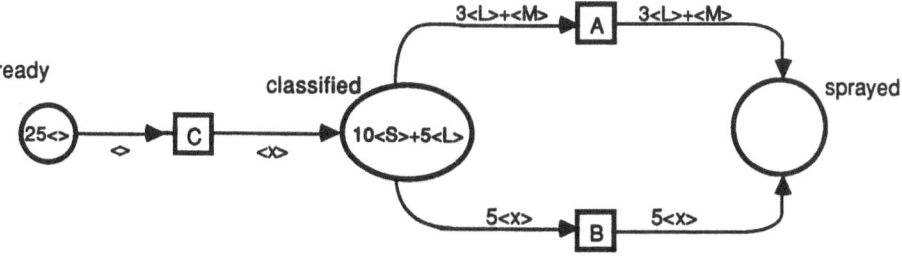

Fig. 5.19

Specification 5.2:

1. This plan describes the painting of forty car bodies.
2. The car bodies will be classified as either "small", or "medium" or "large".
3. Automatic line A sprays three "large" and a "medium" car per run.
4. Automatic line B sprays five cars per run, of any sort.
5. Painted bodies will be collected in one place.

In a PrT net, a finite set of constants — called *colors* — is associated with each S- and T-element. We assume that following colors are associated with the elements of PrT net 5.19:

element	colors
ready	<>
classified	S, M, L
sprayed	S, M, L
C	S, M, L
A	—
B	S, M, L

- S-element "ready" is marked by 25 *tokens* — here represented by symbol "25<>". This marking represents 25 car bodies ready to be painted.
- "ready" is the only input element of transition C. Since arc (ready, C) is labelled "<>", transition C is *enabled* 25 times. Since {s, M, L} is its color set, C can either *occur for color* $x = S$, or for color $x = M$, or for color $x = L$.
  If C occurs for $x = M$, one *tuple* <M> will be added to the marking of S-element "classified"; if C occurs for $x = S$, one tuple <S> will be added to the marking of S-element "classified", and so on.
  Transition C represents the operation of classifying a car body either as "S" (small), or as "M" (medium), or as "L" (large).
- Ten car bodies were already classified "S", and five "L", as shown by the actual marking of S-element "classified".
- Transition A represents spraying on line A. This transition is not enabled: one tuple <M> is missing on S-element "classified".

- Transition B represents spraying on line B, and is actually enabled three times: once for $x = L$, and twice for $x = S$.

Now let:

K	be a finite non-empty set whose elements are called *constants*,
K^m	be the set of *m-tuples* over set K, for $m = 0, 1, 2, \ldots$,
<>	denote the zero-tuple — also called the *token*,
X	be a finite set of variables ranging over K,
X^n	be the set of *n-tuples* over set X, with $n = 1, 2, \ldots$
$M(K^m)$	denote the set of *multisets* over K^m,
$M(X^n)$	denote the set of *multisets* over X^n.

A *predicate/transition net* — *PrT net* — is then a sextuple $N = (S, T, F, C, L, M_0)$ such that:

(i) (S, T, F) is a net where:

S-elements represent predicates over the actual state of resources, and are called *predicates;*

T-elements represent operations which may be performed consuming/producing changing resources, and are called *transitions;*

the flow relation F represents either resource consumption or resource production. Its elements are the *arcs* of the net.

(S, T, F) is called the *underlying net*.

(ii) $C: S \cup T \rightarrow \cup_n \mathbb{P}(K^n)$

is a mapping which associates a set of constant tuples of specific length with each element x of N — predicate or transition.

C is called the *color function* of the net, and $C(x)$ the *color set* of element x.

The common length $A(x)$ of the tuples belonging to the color set of net element x is called the *arity* of x.

(iii) $L: F \rightarrow \cup_n \{ M(K^n) \cup M(X^n) \}$

where n ranges over the set of the arities of elements of S, and

(*) $\forall (a, b) \in F: \quad L(a, b) \in M(K^{A(p)}) \cup M(X^{A(p)})$ with $p \in \{a, b\} \cap S$

Given that $t \in \{a, b\} \cap T$, we assume that exactly $A(t)$ distinct variables appear in the labels of the arcs adjacent to t, and that such variables are set in one-to-one corrispondence to the components of the colors of t.

L is called the *arc labeling of the net*, and $L(a, b)$ the *label* of arc (a, b).

Condition (*) means that the label of an arc is a multiset of tuples whose length equals the arity of the predicate belonging to this arc. We therefore also speak of $A(p)$ as of the arity

(iv) Every mapping $M: S \to \bigcup_n M(K^n)$ such that

$\forall p \in S: M(p) \in M(C(p))$

is called a *marking of the net.*

$M(p)$ is a multiset of constant $A(p)$-tuples, called the *marking of predicate* p.

$M(p)$ represents the multiset of resource units which actually belong to the extension of predicate p.

M_0 is a special marking, called the *initial marking* of the net.

$M_0(p)$ is the multiset of resource units initially belonging to the extension of p.

(v) The following *transition rule* is defined.

Let k be a color of transition t, and $L_k(x, y)$ denote the multiset of constant tuples obtained by substituting the variables of arc label $L(x, y)$ with the corresponding components of k. We set $L_k(x, y) = 0$ for all $(x, y) \notin F$.

Transition t is said to be *enabled for color* k *at marking* M, if

$\forall p \in \bullet t:$ $L_k(p, t) \leq M(p),$ and

$\forall p \in t \bullet:$ $L_k(t, p) \in M(C(p)).$

Enabled transitions *may occur*, or *be activated*.

The occurrence of transition t changes the actual marking M into the marking M' defined by:

$\forall p \in S:$ $M'(p) = M(p) - L_k(p, t) + L_k(t, p).$

Transitions t', ... , t'' are *concurrently enabled at marking* M if each of them is enabled for a certain color at M, and the occurrence of the one does not change the enabling of the others.

According to this definition, a transition may be multiply enabled for one color, and concurrently enabled for different colors.

Every predicate/transition net can be translated into a condition/event net, in general of much larger size. The reason for this incremented size is clear: such a translation requires that predicates are unfolded into various conditions and transitions into various events.

But also the smaller size of Pr/T nets has its drawback. The causal structure of the plan is no longer evident, so that conflicts are more difficult to detect and state-changes more difficult to analyze.

The cornerstone of predicate/transition net analysis is the reachability graph. Owing to its combinatorial nature, both its construction and analysis require computer support for nets of realistic size. When the net size and the available computation tool allow the construction of this graph, conflicts, deadlocks, traps and loops are easily detected. Also, liveness questions can be answered, reachable states and paths between them can be identified, and causal dependencies between processes investigated.

The calculation and interpretation of PrT net invariants is, at the present, still problematic. The existing results appear to be more of theoretical interest than of practical use.

6. Net Plan Executions

The next two chapters are concerned with the execution of Petri net plans. Monitoring execution steps, representing executions, making decisions over planned alternatives, is supported in a natural way by such plans.

Indeed, it is is a very special feature of net plans that both execution states and their changes can be *directly* represented *on the plan*. This allows *monitoring* the dynamics of plan executions as "token games" played on the "net board". Token games, involve activating enabled transitions, either by an execution supervisor or randomly; as a result of such activations, token are moved and new transitions become activated. Monitoring has proven helpful in order to prefigure planned effects of several execution steps, and that way supports "naïve" choice among execution alternatives.

The global representation of partial or complete plan executions is provided by *execution nets,* which we shall introduce below. Such representations are used to compare and evaluate execution alternatives, or for make records of actual executions.

Controlling plan executions involves making "strategic" choices in execution states allowing for different out moves, and requires a clear-cut design of choices. *Control nets* are place/transition nets defined with the issue in mind of developing plans suitable for this sort of decision making. We will introduce them in the last section of this chapter in preparation for the subject matter of Chapter 7: multicriteria decision making over planned execution alternatives.

6.1 Representing Net Plan Executions

Consider a place/transition net plan: reachable markings represent possible states of plan executions, steps represent changes from one such state to another. In a plan execution markings may be reached repeatedly and steps be repeated. Different instances of the same marking or step cannot be distinguished in the net plan, but an execution supervisor can and will distinguish them.

Representation of *plan executions from the viewpoint of a supervisor* is necessary for evaluating and comparing alternative executions, but is also useful for documentation purposes. This section is devoted to the introduction of *execution nets*, a graphic language for representing executions of place/transition net plans from this standpoint. The graphics of execution nets may appear somewhat redundant, but the benefit of their graphical plenitude is a transparent representation of execution mechanics: executed operations are recorded together with available, consumed, produced resources and leftovers. Like every project planning language — graphical or not — execution nets require computer support. A good graphics editor [FJ] will make their development easy and application useful.

For a supervisor, each achieved execution state is a "new" state: states have individuality, and so have state transitions. In place/transition net plans, execution states are markings — that is, sets of places marked by natural numbers. State transitions are steps: multisets of concurrent transition occurrences — multisets, because transitions may fire concurrently. Plan executions are sequences of enabled steps.

Execution nets may be informally described as nets drawn by a supervisor of a Petri net plan who records marking and steps during a specific execution of a given plan. He will draw a circle for each actual S-element holding, a box for each actually occurring T-element, directed arcs for expressing causal connections between circles and boxes.

Our supervisor will get a cycle-free net with some inputless S-elements — representing the initial state of the plan execution — and some outputless S-elements — representing the ending state.

Execution nets have T-graph character but for the allowed inputless and outputless S-elements. Their definition explicitly refers to the execution of a specific plan.

Execution nets mirror the causal structure of the original plan. They may be seen as net records of subsequently available resources and executed operations. We can trace that state Y was reached from state X via step σ, but we can also see which components of states X and Y were affected by step σ, and how.

Execution Nets

Let $N = (S_N, T_N, F_N, W, M_0)$ be a place/transition net and $\sigma = (T_1, \ldots, T_{h-1})$ a sequence of enabled steps, starting at the reachable marking M_1.

As we observed, steps are multisets over T. Therefore, for $i = 1, 2, \ldots, h-1$ let
$$T_i = m_{i1} t_{i1} \oplus m_{i2} t_{i2} \oplus \ldots \oplus m_{in_i} t_{in_i}.$$

Also, assume that
$$M_1 [T_1 > M_2 [T_2 > M_3 \ldots\ldots M_{h-1} [T_{h-1} > M_h.$$

We say that net $E = (S, T, F)$ is an *execution net of* N *starting at marking* M_1 — or, that E represents an *execution of the plan* represented by N — if E was constructed by the following procedure:

- Set $S = \emptyset$, $T = \emptyset$ and $F = \emptyset$;
- for each $s \in S_N$ such that $M_1(s) = m \neq 0$:

  create m distinct S-elements $s \in E$, all of them labelled s;
- for $i = 1$ to $i = h - 1$:

  [create n_i T-elements of E , respectively labelled $t_{i1}, \ldots\ldots, t_{in_i}$;

  for $\forall t_{ij} \in T_i$:

  [if $\exists s \in S_N$: $W(s, t_{ij}) = -n < 0$ define n distinct pairs $(s, t_{ij}) \in S \times T$ to belong to flow relation F;

  if $\exists s \in S_N$: $W(t_{ij}, s) = n > 0$ then

  create n new S-elements $s \in E$, all of them labelled s, and define n distinct ordered pairs $(t_{ij}, s) \in T \times S$ to belong to the flow relation F]] ;
- end.

By construction, execution nets may contain more copies of the same S- or T-element. Sets S and T will then be proper multisets. For any execution net $E = (S, T, F)$ the following holds true:
$$\forall s \in S: \quad /\!/ \cdot s /\!/ \leq 1 \quad \wedge \quad /\!/ s \cdot /\!/ \leq 1$$
$$\forall x, y \in S \cup T: \quad (x, y) \in F^+ \rightarrow (y, x) \notin F^+$$
$$\neg \exists t \in T: \quad /\!/ \cdot t /\!/ = 0 \quad \vee \quad /\!/ s \cdot /\!/ = 0.$$

The place/transition net of Fig. 5.6 represents a plan which can be executed in infinitely many different ways. Figure 6.1 shows two different execution nets of that PT net.

Plan executions start at a certain state, and may end at any other state. Usually, several different executions lead from state X to state Y. Some PT net plans have *maximal executions* — executions with a necessary end — other have not. Maximal executions starting at the initial marking represent complete plan executions.

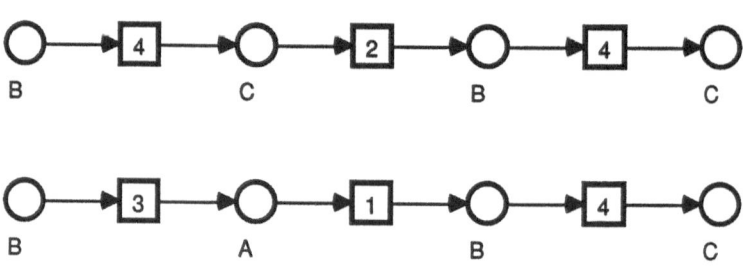

Fig. 6.1

PT net 5.6 has no maximal execution: each execution net can be extended, both backward and forward. Every sequence of enabled steps is a proper subsequence of another sequence of enabled steps.

The net in Fig. 6.2 represents an execution of plan 5.1 interpreted as a PT net with $M_0 = 0$. This execution may be described as follows:

- In the initial state: development of modules B, D, E ready to go, and modules A and C already developed;
- then module B is developed;
- then modules A and B are assembled;
- then assembly AB is assembled with module C;
- end.

This execution cannot be extended forward: it is *forward maximal*.

Isolated S-elements represent resources which are available during the represented execution, but are not affected by it. Non-isolated S-elements represent resources which are produced or consumed during execution; T-elements stand for fulfilled operations.

Inputless S-elements represent resources which are available when the execution starts, *outputless S-elements* represent resources which are available after its ending.

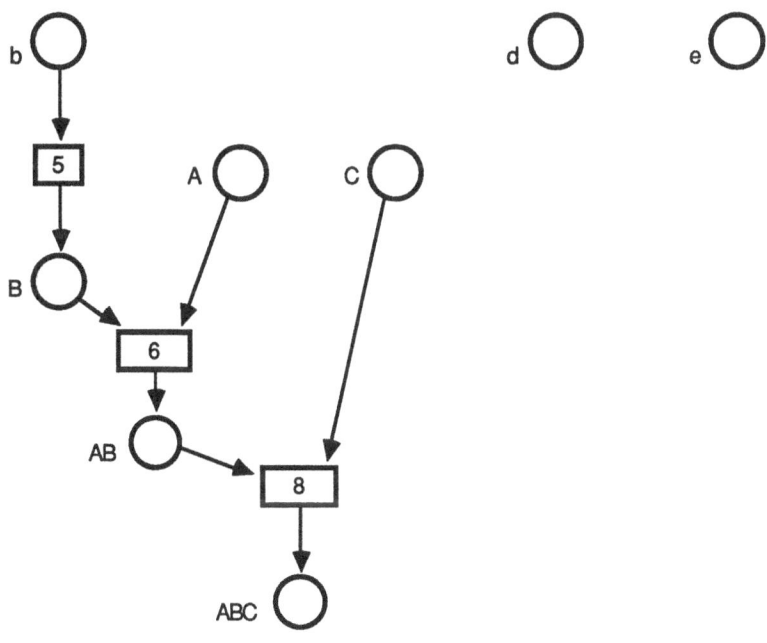

Fig. 6.2

The net in Fig. 6.3 represents a *maximal execution* of PT net 5.1 as it cannot be extended either backward or forward. This execution corresponds to the sequence of enabled steps

$$M_1 [S > M_2 [10\,B \oplus 10\,C > M_3 [D > M_4$$

Other maximal executions of net 5.1 exist.

In Fig. 6.4 a partial execution of the same net is shown which corresponds to the single step

$$M [A \oplus B \oplus 2C > M'$$

where

$$M(b0) = 0, \; M(b1) = 8, \; M(b2) = 7, \; M(b3) = 22, \; M(b4) = 3, \; M(bd) = 0;$$
$$M'(b0) = 0, \; M'(b1) = 0, \; M'(b2) = 5, \; M'(b3) = 30, \; M'(b4) = 5, \; M'(bd) = 0.$$

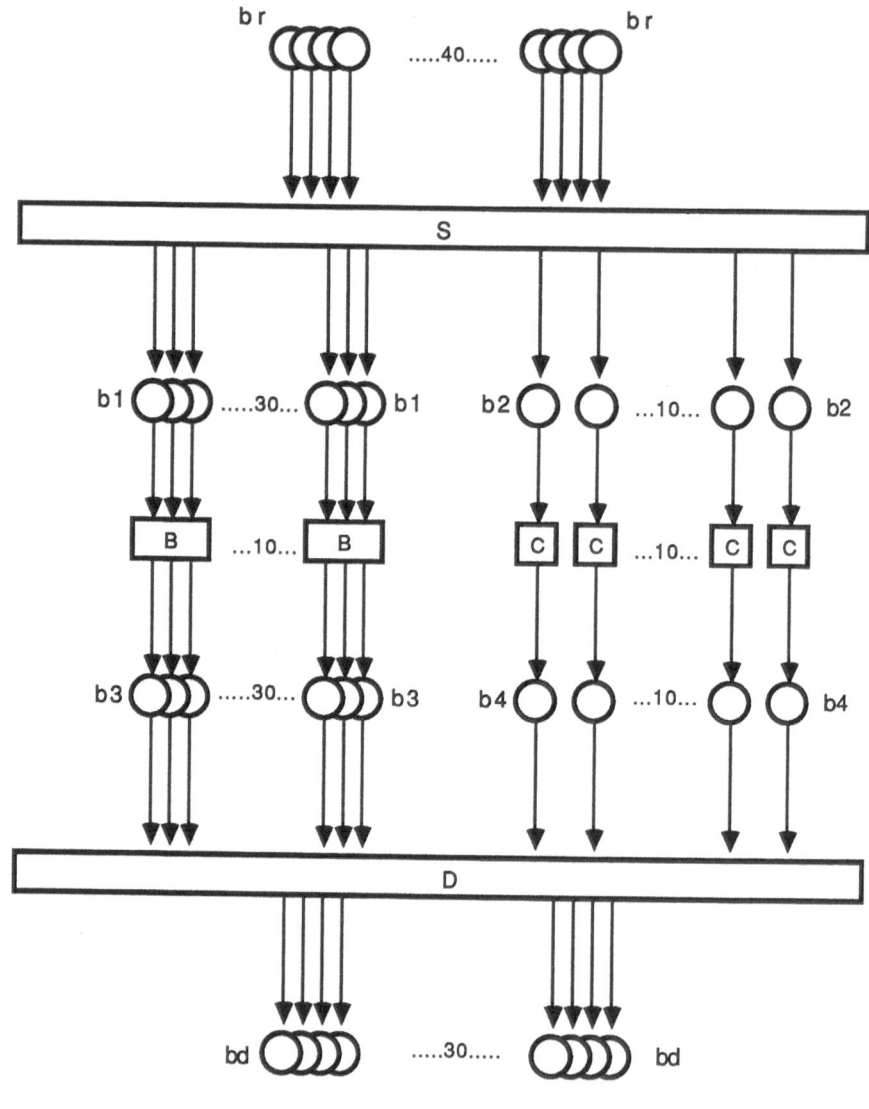

Fig. 6.3

Figures 6.5 and 6.6 show two different executions of the considered plan, both starting at marking

$$M(b0) = 0, \quad M(b1) = 15, \quad M(b2) = 7, \quad M(b3) = 15, \quad M(b4) = 3, \quad M(bd) = 0$$

and ending at the marking

$$M'(b0) = 0, \quad M'(b1) = 0, \quad M'(b2) = 5, \quad M'(b3) = 30, \quad M'(b4) = 5, \quad M'(bd) = 0.$$

In execution nets, maximal place sets such that no two places lie on the same path represent specific instances of net plan markings.

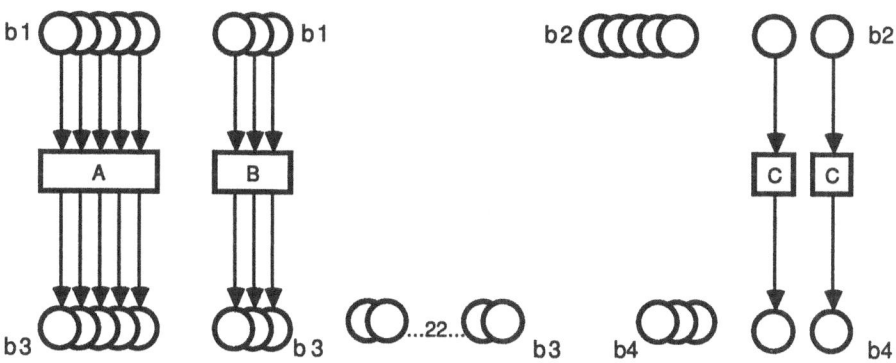

Fig. 6.4

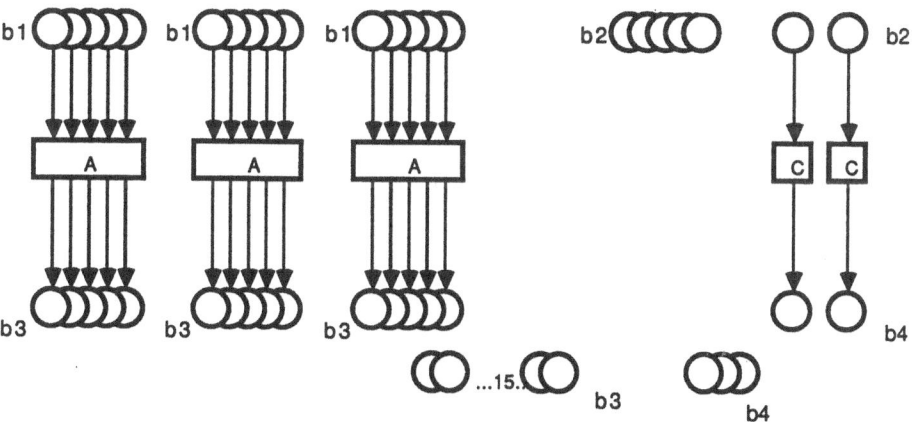

Fig. 6.5

Execution nets are usually disconnected. Connected subnets represent self-sufficient execution fragments

Resources produced by operations of a fragment are never consumed by operations of another fragment. Vice versa, resources consumed by operations of a fragment are never produced by operations of another fragment.

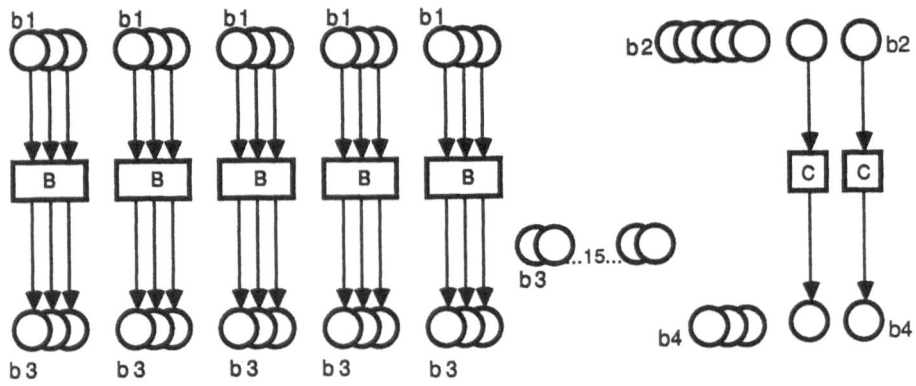

Fig. 6.6

6.2 Figuring Out Execution Alternatives: Control Nets

Execution nets are not suitable for making decisions about plan execution alternatives because the combinatorial generation of all execution nets starting at a given marking is by nature exponentially complex. This is true even though steps are represented compactly — that is, without distinguishing the possible occurrence orderings of the transitions within a step.

If execution control is to include making choices in regard to different allowed "moves", an appropriate type of plan must be used.

Control nets are place/transition nets defined with the issue in mind of how to control decisions. In control nets, planned execution alternatives are by construction organized in clear-cut sets. The choice between such alternatives is controlled by one decision, and by nothing else: once chosen an alternative cannot be blocked or influenced from outside before completion.

Control net plans may be developed just as well in a top-down as in a bottom-up way, and both procedures respect the guidelines of disciplined planning.

T-Sequence

Fig. 6.7

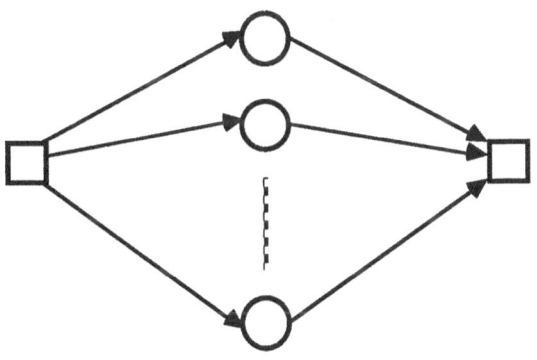

Synchronization

Fig. 6.8

The *top-down development of control nets* is based on *four composition modules* : T-sequence, S-sequence, synchronization and choice.

Figure 6.7 shows the composition module T-sequence; Fig. 6.8 the composition module synchronization. Both of them are T-invariants. T-sequences represent sets of operations to be executed one after the other. Synchronization modules represent simultaneous production and subsequent simultaneous consumption of a set of resources. Isolated transitions will be regarded as T-sequences.

Figure 6.9 shows the composition module S-sequence; Fig. 6.10 the composition module choice. Both of them are S-invariants. S-sequences represent sets of resources which are available one after the other. Choice modules represent resource sharing by alternative operations, both forward and backward. Isolated places will be regarded as S-sequences.

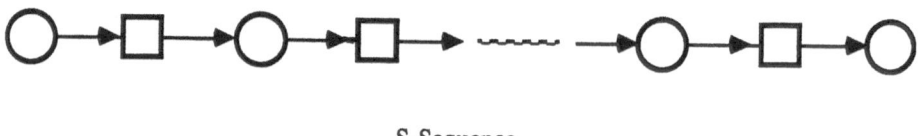

S-Sequence

Fig. 6.9

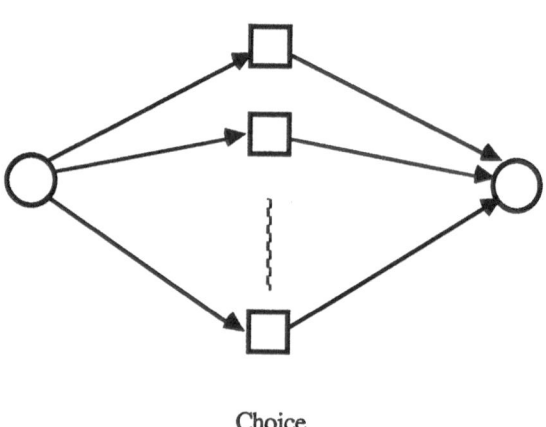

Choice

Fig.6.10

The top-down development of control net plans proceeds as follows:

1. The considered project plan is specified in one of the following ways:

1.1 a set of resources to be sequentially produced and consumed;

1.2 a set of operations to be sequentially carried out ;

1.3 a set of resources produced and consumed simultaneously;

1.4 a set of operations carried out alternatively.

This gives the first, most coarse, control net N representing our plan: a T-sequence, an S-sequence, a synchronization module or a choice module according to 1.1, 1.2, 1.3 or 1.4 being, respectively, the case.

Resources and operations requiring further specification are declared to be *macros* , and so are the places and transitions representing them. (In the sequel, macros will be identified by bold-faced labels.)

If there is no macro node, we are done. Usually, this is not the case, and we proceed.

2.   Macro nodes are separately *specified*.

Macro places will be specified as either sequential or alternative enabling of operations; macro transitions will be specified as sequential or synchronized production/consumption of resources.

3.   For every macro node $\mathbf{m}$ of $\mathbf{N}$, we *substitute* — in accordance with the proper specification — a syntactic module s_m : an S-sequence or a choice module if s_m is a place, a T-sequence or a synchronization module if s_m is a transition.

Places and transition belonging to s_m may be declared to be macros on their turn.

The above substitution is done by setting

$$\cdot\mathbf{m} = \cdot s_m \quad \text{and} \quad \mathbf{m}\cdot = s_m\cdot .$$

We get a new control net N' — by definition a level lower than N. If N' has no macro node, we are done. Otherwise, we substitute N for N', and go back to step 2.

Table 6.1

Transition	Interpretation	Place	Interpretation
S	send to painting	br	unpainted bodies
D	deliver to assembly	**b1**	line-sprayed bodies
		b2	hand-sprayed bodies
		bd	delivered bodies

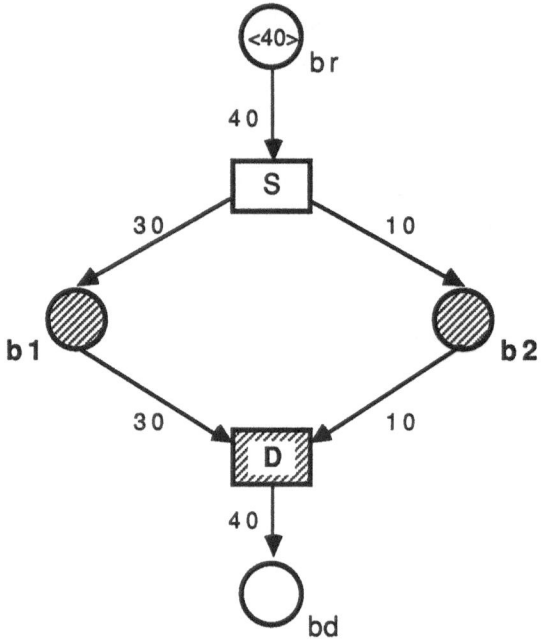

Fig. 6.11

As we have seen, a hierarchy of place/transition nets is step-wise generated by sub-stituting allowed syntactic modules to macro sequences on the basis of ad hoc specifica-tions. We shall illustrate this procedure using a known example: the plan for a car body painting shop of Specification 5.1. For the reader's convenience, we recall that specifica-tion.

Specification 5.1:

1. This plan describes the painting of forty car bodies.
2. Forty car bodies will be sent together to the painting shop.
3. Thirty of them will be automatically sprayed on lines A and B; ten will be hand-sprayed, one body at a time.
4. Line A picks up and sprays five car bodies per run; line B three.
5. Once painted, the forty car bodies will be delivered together to the assembly shop.

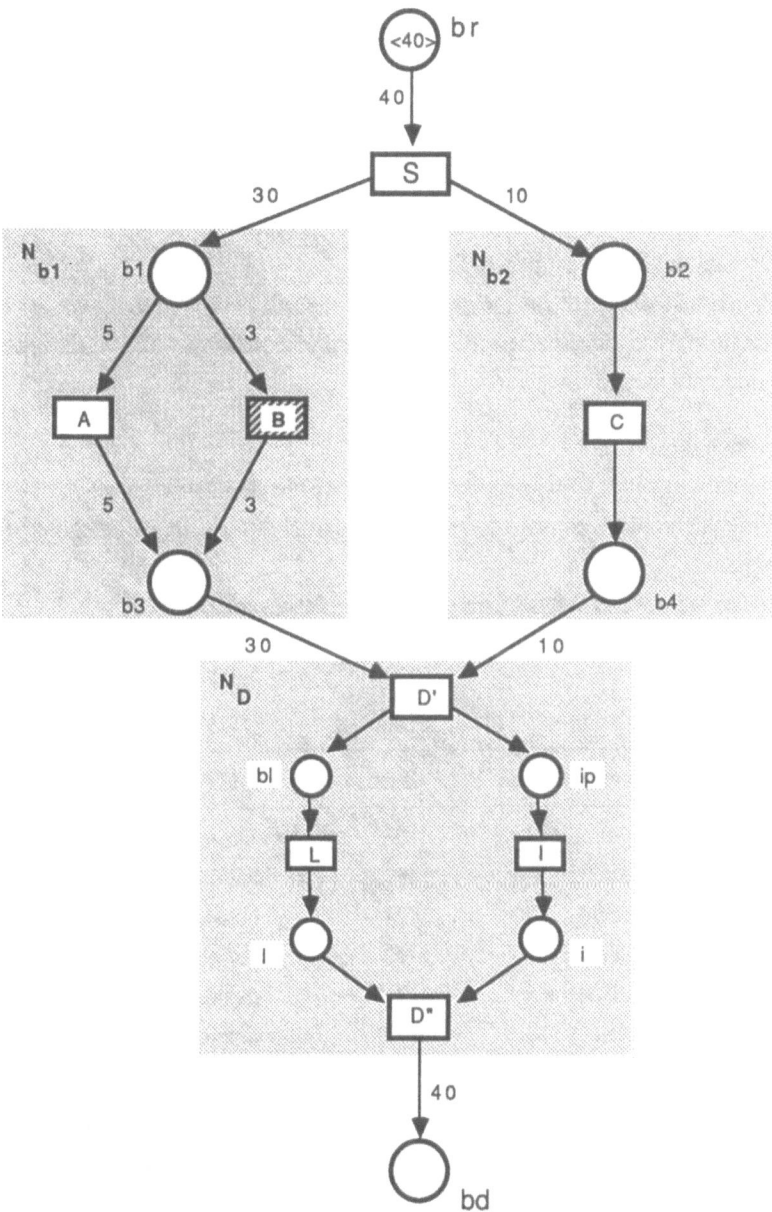

Fig. 6.12

In order to conform to the above directives, we take points 1, 2, 3 and 5 of Specification 5.1 as the first plan specification. That way, the plan is described as a set of resources produced and consumed simultaneously. Figure 6.11 shows the corresponding, first level, control net. Its interpretation is given by Table 6.1. The illustrated marking — 40 tokens in place br — represents forty car bodies ready to be painted. Places **b1** and **b2**, and transition **D** are declared to be macros. Therefore, they are identified by bold-faced labels, and shadowed in the figure.

Figure 6.12 shows the next top-down development step. We took points 3 and 4 of Specification 5.1 as the specifications of macro places **b1** and **b2**. Macro transition **D** will require further specification:

Specification D:
1. The delivery of the forty painted car bodies will be done in one batch.
2. While the bodies are loaded on the delivery truck, the accompanying invoice will be prepared.
3. Bodies and invoice will be carried together to the assembly shop.

Table 6.2

Transition	Interpretation	Place	Interpretation
A	spraying on line A	b3	automatically sprayed bodies
B	spraying on line B	b4	hand-sprayed bodies
C	hand-spraying	bl	bodies to be loaded
D'	prepare delivery	ip	invoice to be processed
D"	accomplish delivery	l	loaded bodies
L	load bodies	i	prepared invoice
I	prepare invoice		

The plan in Fig. 6.12 has been obtained by unfolding the macros **b1**, **b2** and **D**. The interpretation of additional nodes is given in Table 6.2. Place **b1** has been substituted with a choice module in accordance with point 4 of Specification 5.1; place **b2** has been substituted with a sequence as from point 3 — bodies *individually* hand-sprayed. Transition **D** with a synchronization module conforms to Specification D.

We declare the "new" transition **B** to be a macro.

The next step consists in unfolding transition **B**. We assume the following specification to be given

<u>Specification B:</u>
Spraying on line B may be done using paint of color 1, 2, ... , or n.

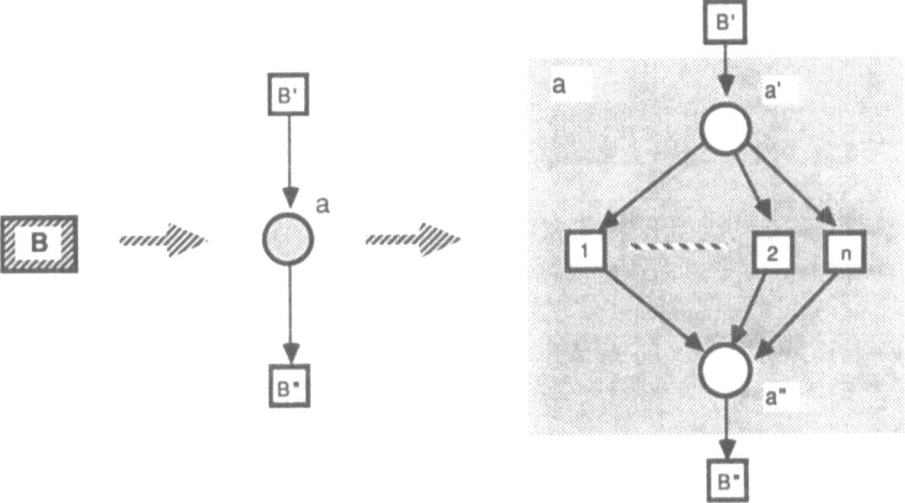

Fig. 6.13

Figure 6.13 shows the control net unfolding of **B**. Transition **B** is first substituted with a T-sequence — the smallest non-trivial T-sequence. The place of the T-sequence is then substituted by means of a choice module. The previous substitution of transition **B** with a T-sequence is necessary in order to set the choice of the spraying color at a level of decision lower than the choice between spraying on line A or on line B.

The interpretation of the "new" nodes in the unfolding of **B** is given in Table 6.3. Since none of them is declared to be a macro, the top-down development procedure of our plan terminates here.

Table 6.3

Transition	Interpretation	Place	Interpretation
B'	set up line for spraying	a'	line ready for spraying
1	spraying in color 1	a"	spraying terminated
...	.....		
n	spraying in color n		
B"	reset line after spraying		

Control nets may also be constructed bottom-up, on the basis of the following definition:

1. Place/transition nets made up of either a transition with one input and one output place, or a place with one input and one output transition, are *control nets*.

    Figure 6.14 represents two such control nets. Notice that any marking and any arc weight is allowed. Figure 6.15 shows shorthands for generic control nets of both the types in Fig. 6.14. *Sources* are vertically shaded, *sinks* horizontally.

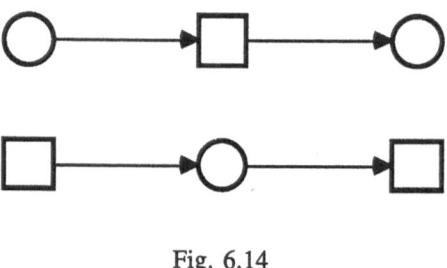

Fig. 6.14

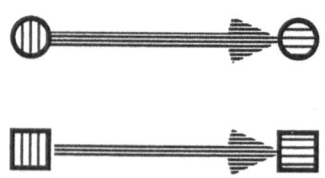

Fig. 6.15

2. *Two composition rules* are given: sequence and parallel composition.
    Sequence : for any two or more control nets, the net obtained by identifying the source of the one with the sink of the other is a control net.
    Figure 6.16 illustrates the application of the composition rule sequence.

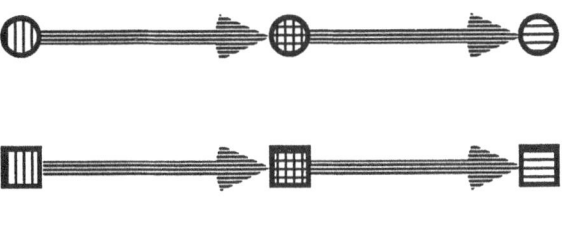

Fig. 6.16

Parallel composition : for any two or more control nets, the net obtained identifying both their sources and their sinks is a control net.
Figure 6.17 illustrates the application of parallel composition.

3. Nothing else is a control net.

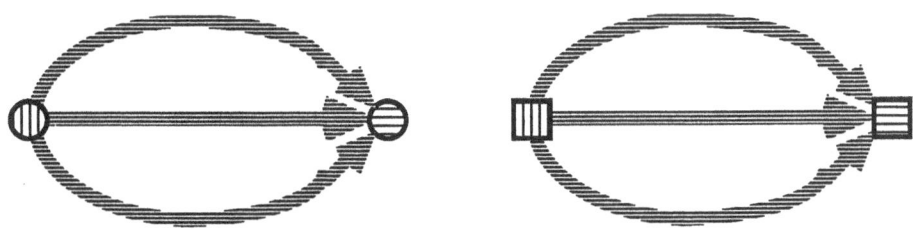

Fig. 6.17

As an instructive exercise, we suggest carrying out both the top-down and the bottom-up development of the control net in Fig. 6.18.
    The syntactic module S-sequence may now be seen as obtained by applying sequence composition to two control nets of the sort above in Fig. 6.14 above. The module

T-sequence, as obtained by applying sequence composition to two control nets of the sort below in the same figure.

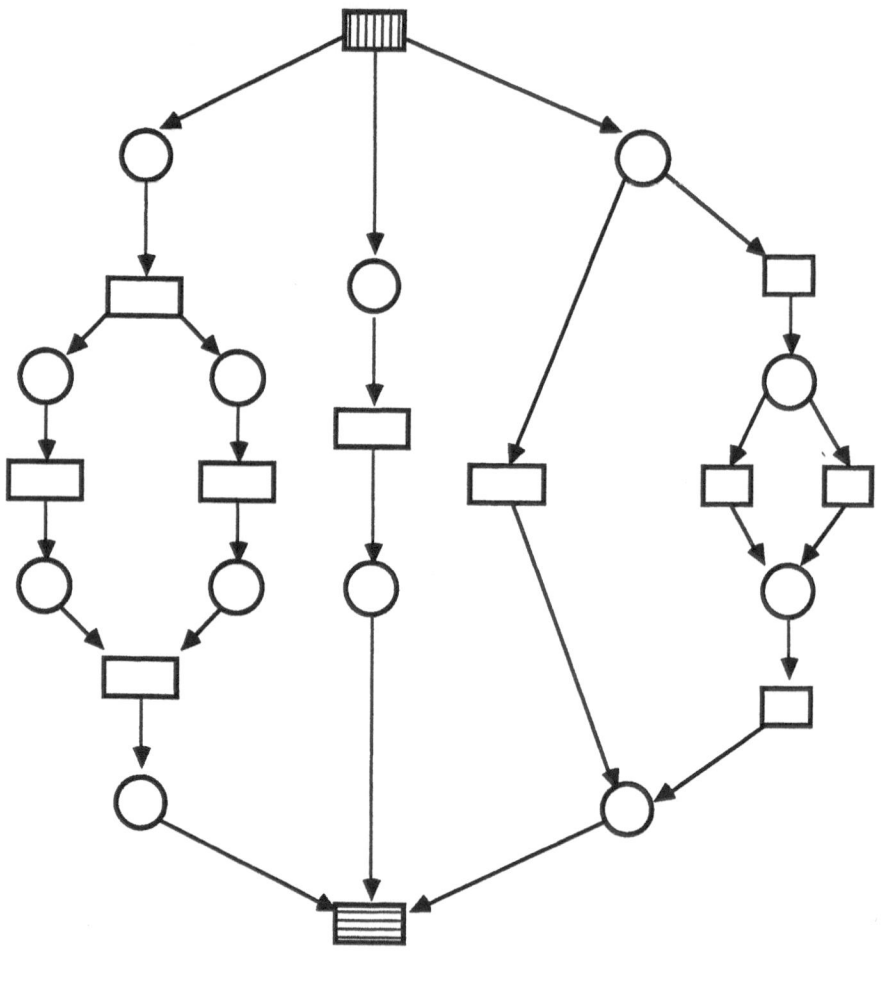

Fig. 6.18

Control nets are constrained in structure. The gain of this limitation is the clear-cut definition of execution alternatives. Decision over execution alternatives arise at some places of choice modules, and only there. Furthermore, every branch out of such a place is guaranteed to represent a choosable alternative.

By the way of contrast, consider PT net 6.19 — not a control net. Two "alternatives" branch from place a. But one of these branches may or may not be choosable, depending on how the choice at place b is resolved. The decision about executing transition 1 or transition 2 is taken in place a *and* in place b. Place a and place b are not the sites of two independent decisions. Rather, set {a, b} is the distributed site of one compound decision.

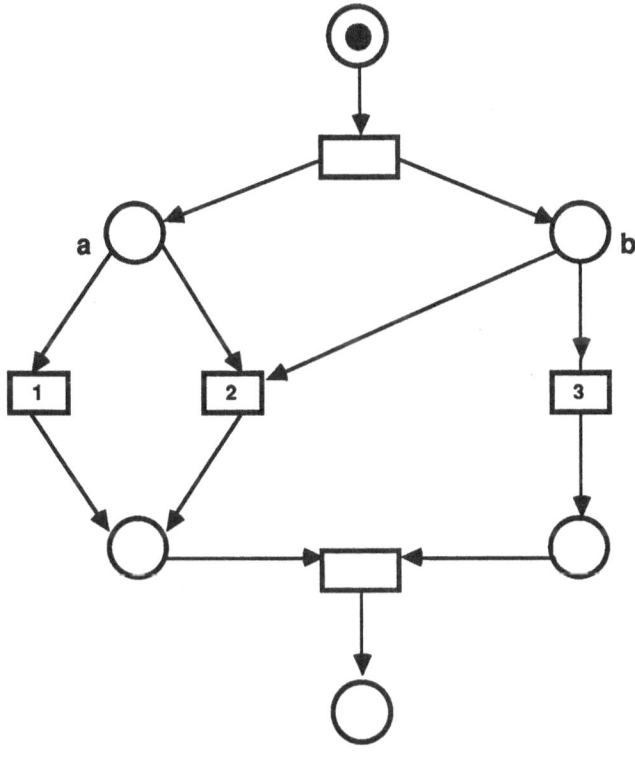

Fig. 6.19

In the next chapter we shall introduce an effective technique for making decisions over execution alternatives. For this purpose we need some additional definition about "choice structure" in control nets.

Here we shall call a *choice schema* any subnet having a source and a sink, both of them places. In top-down developed control nets, the choice schemata are the choice modules and their unfoldings. In bottom-up developed nets, choice schemata result from the parallel composition of control nets whose source and sink are places.

The source of a decision schema is called the *control place*. Control places are state components representing planned request of choice. Choice schemata can be uniquely identified by their control place — at any level of detail. Therefore, we will often denote control nets by their control place label.

Alternatives of choice schemata are T-graph subnets which contain both its source and sink. They can, and will, be uniquely identified as sets of transitions.

If choice schema c' is a subnet of choice schema c, we will say that c' is *nested* in c. Choice schemata not nested in other choice schemata are said to be *primary*. Each choice schema is primary at some plan level.

Alternatives of *primary* choice schemata are either S-sequences, or S-sequences in which some elements have been substituted, by synchronization modules or other sequences, possibly several times. Each alternative is a sequence at some plan level.

The control net obtained by carrying out the development step of Fig. 6.13 on control net 6.12 has two choice schemata: **b1**, identified by node set $\{b1, A, B', a', 1, 2, \ldots, n, a'', B'', b3\}$, and a', identified by node set $\{a', 1, 2, \ldots, n, a''\}$. a' is primary, while **b1** is not. The alternatives of a' are: $\{1\}, \{2\}, \ldots, \{n\}$. The alternatives of **b1** are $\{A\}, \{B', 1, B''\}, \{B', 2, B''\}, \ldots, \{B', n, B''\}$. At the plan level represented in Fig. 6.12, **b1** is primary.

An *alternative* A is said to be *enabled at marking* M if there exists a realizable occurrence sequence made up of all the transitions belonging to A, and only of them, and starting from marking M. Such an occurrence sequence is called an *execution of alternative* A *from marking* M.

Two alternatives are said to be *concurrently enabled* at marking M, if they are both enabled at M, and the execution of the one leaves the other enabled. Alternatives may be multiply enabled.

A *choice schema* is said to be *enabled at marking* M if at least one of its alternatives is.

Given a choice schema s, the execution from a marking M of any combination of concurrently enabled alternatives of s is called an *execution of* s, if:

(i)   at marking M, the control place is the only marked place,

(ii) after that execution, the control place no longer marked.

Let $s_1, s_2, \ldots, s_n$ be the alternatives of choice schema s concurrently enabled at marking M. The multiset

$$m_1 s_1 \oplus m_2 s_2 \oplus \ldots \oplus m_n s_n$$

represents an execution of choice schema s started from marking M, and in which alternative s_1 was carried out m_1 times, alternative s_2 was carried out m_2 times, ... , and alternative s_n was carried out m_n times, *in arbitrary order*.

7. Choosing a Course of Action: Operational Decision Making

Plans allow for alternative executions, and different executions will mostly have a different "degree of desirability", depending on times, costs, failure rates, setup times, etc. Choosing an optimal course of action whenever plans allow for alternative moves is a crucial problem of plan execution supervision. In this chapter we present a methodology for making decisions over alternative executions of plans represented as control nets. We called this methodology *Operational Decision Making* (ODM), because it is suitable for making decisions at the operational level of running a plan.

Suppose that a control net representation of a certain plan is given, and we intend to execute it. An *execution supervisor* will be in charge of activating, sooner or later, all non-conflicting enabled transitions, and of making decisions in choice situations. He will want to make good decisions in the light of his knowledge about planned operations and intended goals. ODM is a methodology to support making these choices. It requires that the valuations that are to govern choosing be expressed in a certain form.

Our supervisor will consider "vague" properties of operations. He will attach *fuzzy attributes* to operations — such as short times, low costs, acceptable failure rates, brief setup times, low labour rates, etc. From operation attributes, he will deduce attributes of the alternative executions — fuzzy too. This done, the supervisor has different multiattribute decision making techniques at his disposal. He may either find optimal alternatives by *intersecting* attributes, or he may apply *outranking techniques* in order to partition the set of alternatives into several preference classes ranked according to a desirability scale defined by the supervisor.

ODM requires plans where alternative courses of action do not interact. Control net plans are by construction such, but ODM can be applied to any plan representation provided the independence of execution alternatives is guaranteed.

As we know, there is associated with each control net a hierarchy of plans. For simplicity of exposition we will assume that all control nets in the next pages are developed top-down. But the techniques under discussion are surely independent of whether plans are generated top-down or bottom-up.

7.1 Execution Alternatives

Control nets allow for alternative executions if, and only if, they contain at least one choice module. The execution of control net plans requires a decision on "what to do" every time more than one alternative is enabled. The supervisor views the control places of such schemata as information requests: "Which course of action should best be taken?" Or: "Which alternative, or combination of alternatives, should next be activated?"

During the resolution of choice schemata — decision making about which alternatives to execute — the marking of the control place stays constant: the supervisor will not activate any transition. This makes sure that the spectrum of possibilities is not modified before the decision is made, in particular that, once made, decisions will still be executable.

Plans with nested choice schemata are handled by resorting to the hierarchy of plans: decisions about choice schemata are always made at the lowest hierarchical level at which they are primary.

Operational decision making requires fuzzy attributes to be associated to operations. The next section is therefore devoted to introducing some rudiments of fuzzy set theory.

7.2 The Formal Representation of Vagueness: Fuzzy Sets

Fuzzy sets were first introduced by L. A. Zadeh in the sixties. After three decades of research, the theory of fuzzy sets now provides the conceptual framework for dealing with problems involving a special sort of uncertainty: the vagueness deriving from the absence of sharply defined criteria of set membership.

The imprecision involved in locutions like "Francis is young", " Helen has dark hair", "Operations A and B have the same duration" is of this sort. An imprecision by far different from the uncertainty arising in stochastic questions, where the uncertainty affects the holding of sharply specified states, or the taking place of sharply specified events.

In classical set theory, an element either belongs or does not belong to a set. In fuzzy set theory, membership of elements in sets is more finely tuned: real numbers between zero and one are used in order to express the membership degree of an element in a fuzzy set.

More formally: given a set E, we call any set of ordered pairs with

$$A = \{ \ (e, \mu(e)) \ | \ e \in E \ \wedge \ \mu(e) \in [0, 1] \ \}$$

a *fuzzy set over* E.

E is called the *support* of A, and $\mu(e)$ the *membership degree* of e in A. We shall write $A = \{e, \mu(e)\}$ instead of $A = \{ \ (e, \mu(e)) \ | \ e \in E \ \}$ when this causes no ambiguity. When $\mu(e) = 0$ we say that element e does not belong to fuzzy set A.

Given any real number $\alpha \in [0, 1]$, we call the subset E^α of E defined as

$$E^\alpha := \{ \ e \in E \ | \ \mu(e) \geq \alpha \ \}$$

the α–*cut* of A.

Set operations are extended to fuzzy sets. The extension is carried out by means of the membership functions:

Given two fuzzy sets $A = \{e, \mu_A(e)\}$ and $B = \{e, \mu_B(e)\}$ with same support E, we say that A *includes* B, and write $A \supseteq B$, if for all $e \in E$

$$\mu_A(e) \geq \mu_B(e);$$

we say that A *strongly includes* B, and write $A \supset B$, if for all $e \in E$

$$\mu_A(e) > \mu_B(e).$$

The *intersection* of the above fuzzy sets A and B is defined as the largest fuzzy set which is contained in both A and B:

$A \cap B := \{ (e, \mu_{A \cap B}(e)) \mid e \in E \wedge \mu_{A \cap B}(e) = \min \{ \mu_A(e), \mu_B(e) \} \}$.

The *union* of the above fuzzy sets **A** and **B** is defined as the smallest fuzzy set which contains both **A** and **B**:

$A \cup B := \{ (e, \mu_{A \cup B}(e)) \mid e \in E \wedge \mu_{A \cup B}(e) = \max \{ \mu_A(e), \mu_B(e) \} \}$.

Let **X** and **Y** be two ordinary sets. We will call every fuzzy set

$$R := \{ ((x, y), \mu_R(x, y)) \mid (x, y) \in X \times Y \}.$$

a *fuzzy relation on* $X \times Y$. Set $X \times Y$ is called the *support of the relation*.

Given two fuzzy relations

$R_1 = \{(x, y), \mu_1(x, y)\}$ on $X \times Y$ and $R_2 = \{(y, z), \mu_2(y, z)\}$ on $Y \times Z$

the fuzzy set

$$R_1 \circ R_2 := \{ ((x, z), \max_{y \in Y} \{ \min \{ \mu_1(x, y), \mu_2(y, z) \} \}) \mid x \in X, z \in Z \}$$

is called the *max-min composition of* R_1 and R_2.

An ordinary relation $R := \{ (x, y) \mid (x, y) \in X \times Y \}$ can be represented as a bipartite digraph the vertex set of which is $X \cup Y$, and whose directed arcs are the pairs $(x, y) \in R$.

A fuzzy relation $R := \{ ((x, y), \mu_R(x, y)) \mid (x, y) \in X \times Y \}$ can be represented as a weighted bipartite digraph. To this end, weights $\mu_R(x, y)$ will be associated with arcs (x, y) in the bipartite digraph representing the associated ordinary relation $R := \{(x, y), \mid (x, y) \in X \times Y \}$.

Graphs weighted by means of the membership degrees of a fuzzy relation are called *fuzzy graphs*.

7.3 Fuzzy Attributes

Let $N = (S, T, F, W, M_0)$ be a control net, and $c = (S', T', F', W', M')$ a primary choice schema of it. We assume that the plan represented by net N is being executed, and that c is actually enabled. In order to execute the plan, the supervisor has to make a decision about which alternative will next be activated.

Attributes over Operations

We assume that the execution supervisor is able to associate a finite set of fuzzy attributes with T_c, the transition set of c.

A *fuzzy attribute over operations* is a fuzzy set $A = \{t, \mu(t)\}$ with support T_c. The degree of membership $\mu(t)$ of transition t is to be interpreted as the degree to which operation t exhibits attribute A.

Short durations, low costs, small failure rates, etc. are examples of properties which may well be expressed as fuzzy attributes. Consider, for instance, the choice schema c in Fig. 7.1. The property "low cost" could be specified by the fuzzy set
$$C = \{t, \mu_C(t)\} \quad \text{with}$$
$$\mu_C(t) = \{(1, 0.9), (2, 0.4), (3, 1.0), (4, 0.2), (5, 0.8)\},$$
while property "short duration" could be specified by fuzzy set
$$D = \{t, \mu_D(t)\} \quad \text{with}$$
$$\mu_D(t) = \{(1, 0.2), (2, 0.8), (3, 0.0), (4, 0.6), (5, 0.4)\}.$$

These two attributes mean: "The degree to which operation 1 has the attribute Low Cost is 0.9; the degree to which it has the attribute Short Duration is 0.2", etc.

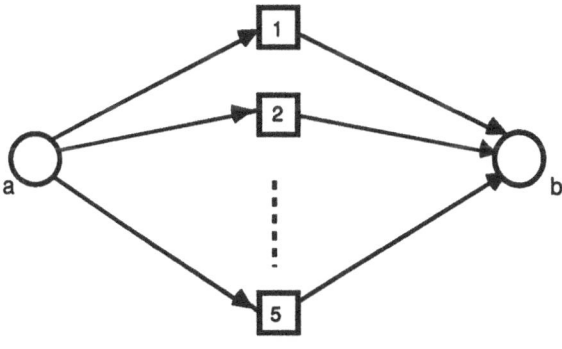

Fig. 7.1

We shall be concerned with decisions between alternative executions of choice schemata. Therefore, attributes must be associated with the executions. We do this on the basis of attributes associated with the involved operations.

Attributes over Alternatives

Executions of choice schemata generally involve the activation of several transitions. Extending fuzzy attributes of operations one by one to fuzzy attributes of executions is a rather natural idea. However, to do so requires taking the meaning of the individual attributes into account. For instance, the aggregation of time related attributes must be different from the aggregation of cost related ones. Indeed, while it is natural to view the joint cost of two *parallel* activities as the sum of their costs, their joint duration will naturally be interpreted as the union of the corresponding time intervals.

Alternatives of primary choice schemata are either S-sequences or unfoldings of S-sequences. In any case, they are are made up of sequences and of synchronization modules. Hence, the extension of an attribute over operations to an attribute over executions requires *two operators*, one for aggregating the membership degrees of transitions belonging to sequences, and one for aggregating the membership degrees of transitions belonging to synchronization modules. These two operators must be specified by the supervisor for *every* operation attribute, in accordance with our definition below. The specification of these two operators expresses the supervisor's beliefs about how the considered attribute spreads from operations to execution alternatives.

Consider a primary choice schema c, and a fuzzy attribute $A = \{t, \mu\ (t)\}$, defined over its set of transitions. Let $C = \{c_1, c_2, \dots, c_n\}$ be the set of alternatives of c.

We define a *fuzzy attribute over the alternatives* of c to be a fuzzy set $A' = \{c_i, \mu'(c_i)\}$ with support C which membership function is computed on the basis of a fuzzy operation attribute A by means of two operators assigned by the supervisor: operator § for aggregating the membership degrees of operations belonging to sequences, and operator # for aggregating the membership degrees of operations belonging to synchronization schemata. Operators § and # must be specified so that:
- for every S or T-sequence s with transition set $\{t_1, t_2, \dots, t_n\}$
$$\S(s) := \S(\mu(t_1), \mu(t_2), \dots, \mu(t_n)) \in [0,1] ;$$
- for every synchronization schema y with transition set $\{t_1, t_2, \dots, t_n\}$
$$\#(y) := \#(\mu(t_1), \mu(t_2), \dots, \mu(t_n)) \in [0,1].$$

The membership degree $\mu'(c_i)$ of alternative c_i for the fuzzy attribute over alternatives A' is then obtained in the following way:
1. If c_i is an S-sequence — remember that isolated places are such — we set
$\mu'(c_i) := \S(c_i)$ and terminate.

2. If not, we run bottom-up through the hierarchy of plans, which we assume indexed
    by $1, 2, \ldots, n$, and carry out the following computations:
    Set $\mu_n(t) := \mu(t)$ for all transitions t, and $h := n$;

    REPEAT

    *   decrement h by 1
    *   for each macro place **p** of the level h plan which is substituted by an S-sequence x
        in the level $h + 1$ plan, set
        $$\mu_h(p) := \S(x) \,;$$
    *   for each macro transition **t** of the level h plan which is substituted by a T-sequence
        y in the level $h + 1$ plan, set
        $$\mu_h(t) := \S(y) \,;$$
    *   for each macro transition **t** of the level h plan which is substituted by synchroniza-
        tion module **w** in the level $h + 1$ plan, set
        $$\mu_h(t) := \# (w) \,;$$
    *   for all places p and transitions t for which $\mu_{h+1}(p)$ or respectively $\mu_{h+1}(t)$ was
        defined, set
        $$\mu_h(p) := \mu_{h+1}(p) \quad \text{and} \quad \mu_h(t) := \mu_{h+1}(t)$$
    UNTIL in the level h plan c_j contains exactly one transition t';
3. Set $\mu'(c_i) := \mu_h(t')$, and terminate.

Attributes over Executions

Fuzzy attributes over alternatives lead naturally to fuzzy attributes over executions.

Executions of choice schemata are, by definition, executions of multisets of concurrently enabled alternatives. If $c_1, c_2, \ldots, c_n$ are the actually enabled alternatives of
choice schema c, multiset
$$e := m_1 c_1 \oplus m_2 c_2 \oplus \ldots \oplus m_n c_n$$
represents the execution of c consisting in executing — in any order and for all i — m_i
times alternative c_i.

Let E denote the actual execution set of c. The *fuzzy attribute over the executions* of
c is the fuzzy set $A'' = \{e, \mu''(e)\}$, with support E and with the membership function
$$\mu''(e) := [m_1 \, \mu'(c_1) + m_2 \, \mu'(c_2) + \ldots + m_n \, \mu'(c_n)] / \Sigma_i \, m_i$$
with
$$\Sigma_i \, m_i = M(c').$$

The membership degree of execution **e** in the fuzzy attribute over executions is the weighted average of the membership degrees of the alternatives constituting **e** in the corresponding fuzzy attribute over alternatives. The membership degree of **e** in fuzzy set **A"** is a linear function of the membership degree of c_i in fuzzy set **A'**.

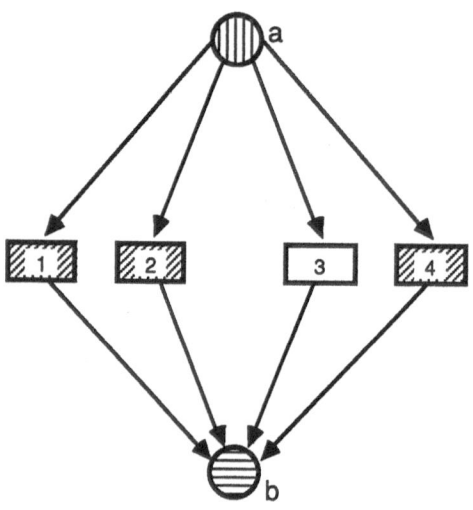

Plan N: Level 1

Fig. 7.2

As an example, we shall consider two execution attributes of plan **N** represented in Fig. 7.5. The figure is the result of top-down expanding Fig. 7.2 in three steps. The intermediate results are shown in Figs. 7.3 and 7.4. The level 1 plan is represented by a choice schema with three macro transitions: **1, 2,** and **4**. The level 2 plan contains three macro places: **c, d,** and **e**. The level 3 plan contains macro place **f** and macro transition **1C**.

Let $\mathbf{D} = \{t, \mu_A(t)\}$ and $\mathbf{K} = \{t, \mu_B(t)\}$ be two fuzzy attributes — respectively called Short Duration and Low Cost — over the transitions of choice schema 7.5. The support of **D** and **K** is, by definition, the transition set of net 7.5. That is:

$$T = \{1A, 1CA, 1CB, 1D, 1B, 2A, 2C, 2B, 3, 4A, 4C, 4B\},$$

Assume that the membership functions of **D** and **K** are:

Transition	$\mu_D(t)$	$\mu_K(t)$
1A	0.9	0.7
1CA	0.3	0.8
1CB	0.4	0.8
1D	0.5	0.5
1B	0.9	0.7
2A	0.5	0.5
2C	0.8	0.4
2B	0.5	0.5
3	0.8	0.7
4A	0.4	0.4
4C	0.6	0.3
4B	0.4	0.6

In order to compute fuzzy attributes Short Duration and Low Cost over the set of alternatives, the aggregating operators § and # must be specified. This will express our beliefs about the aggregating behavior of short durations and low costs. Assume here that operators § and # are specified by:

1. if x is a sequence, and $e_1, e_2, \ldots, e_n$ its transitions and macro places:
$$\S_D(x) := [\Sigma_{i=1, 2, \ldots, n} \ \mu_D(e_i)] / n ;$$
$$\S_K(x) := [\Sigma_{i=1, 2, \ldots, n} \ \mu_K(e_i)] / n ;$$

2. if y is a synchronization module, and $x_1, x_2, \ldots, x_m$ its maximal sequences:
$$\#_D(y) := \min_{i=1, 2, \ldots, m} \ \S_D(x_i) ;$$
$$\#_K(y) := [\Sigma_{i=1, 2, \ldots, m} \ \S_K(x_i)] / m .$$

This specification of operators § and # embodies the view that the planned costs will sum up, and so will the durations of sequential operations, while the duration of bundles of synchronized processes will coincide with the duration of the slowest process.

Now we are prepared to compute the fuzzy attributes Short Duration and Low Cost over alternatives. We will denote them as $D' = \{c_i, \mu'_D(c_i)\}$ and $K' = \{c_i, \mu'_K(c_i)\}$, respectively.

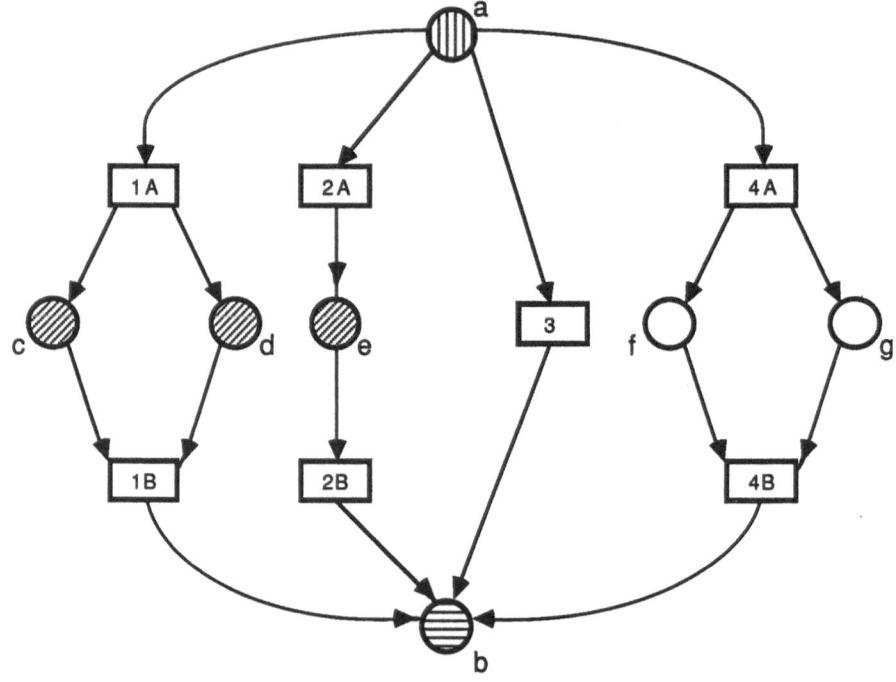

Plan N: Level 2

Fig. 7.3

The alternatives of choice schema 7.5 are (denoted as sets of transitions):
$$c_1 = \{1A, 1CA, 1CB, 1D, 1B\}, \quad c_2 = \{2A, 2C, 2B\},$$
$$c_3 = \{3\}, \quad c_4 = \{4A, 4C, 4B\}.$$
c_2 and c_3 are sequences, and hence:
$$\mu'_D(c_2) = (0.5 + 0.8 + 0.5)/3 = 0.6 \quad \text{and} \quad \mu'_D(c_3) = 0.8,$$
$$\mu'_K(c_2) = (0.5 + 0.4 + 0.5)/3 = 0.46 \quad \text{and} \quad \mu'_K(c_3) = 0.1.$$
c_1 is not a sequence, and $\mu'_D(c_1)$ and $\mu'_K(c_1)$ are computed on the basis of the top-down hierarchy of plans. Choice schema 7.5 is the level 4 plan.

Let us first apply the above algorithm to the computation of $\mu'_D(c_1)$. We set:
$$h := 4,$$
$$\mu'_{D4}(1A) = \mu_D(1A) = 0.9, \quad \mu'_{D4}(1B) = \mu_D(1B) = 0.9, \quad \mu'_{D4}(1D) = \mu_D(1D) = 0.5,$$
$$\mu'_{D4}(1CA) = \mu_D(1CA) = 0.3, \quad \mu'_{D4}(1CB) = \mu_D(1CB) = 0.4.$$

We then go back to level 3. Here, the only macro element is transition **1C**. We get:

$$h := 3,$$

$$\mu'_{D3}(1A) = 0.9, \quad \mu'_{D3}(1B) = 0.9, \quad \mu'_{D3}(1D) = 0.5,$$

$$\mu'_{D3}(1C) = \#_D(\{1CA, 1CB\}) = \min \{ \S_D(\{1CA, 1CB\}), \S_D(\{1CA, 1CB\}) \} =$$

$$= \min \{ (0.3 + 0.4)/2, (0.3 + 0.4)/2 \} = 0.35.$$

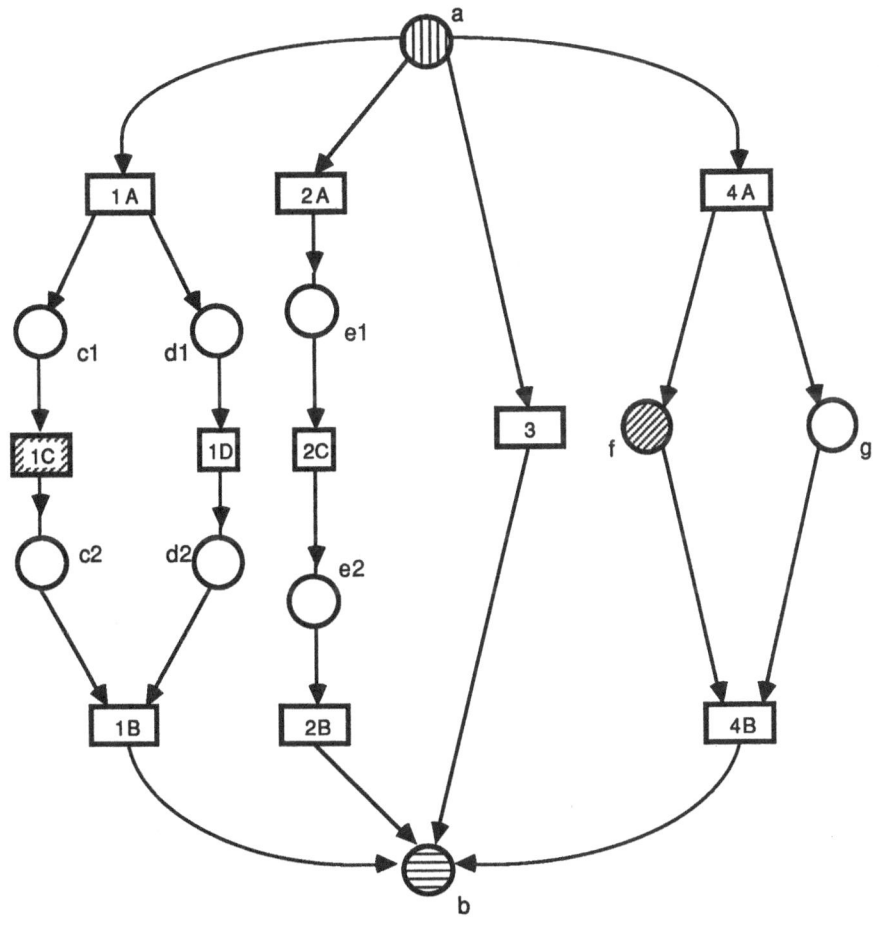

Plan N: Level 3

Fig. 7.4

Now we step back to level 2, where places **c** and **d** happen to be macros. We get:

$$h := 2,$$

$$\mu'_{D2}(1A) = 0.9, \quad \mu'_{D2}(1B) = 0.9, \quad \mu'_{D2}(1D) = 0.5, \quad \mu'_{D2}(1C) = 0.35,$$

$$\mu'_{D2}(c) = \S_D(\{1C\}) = 0.35, \quad \mu'_{D2}(d) = \S_D(\{1D\}) = 0.5 .$$

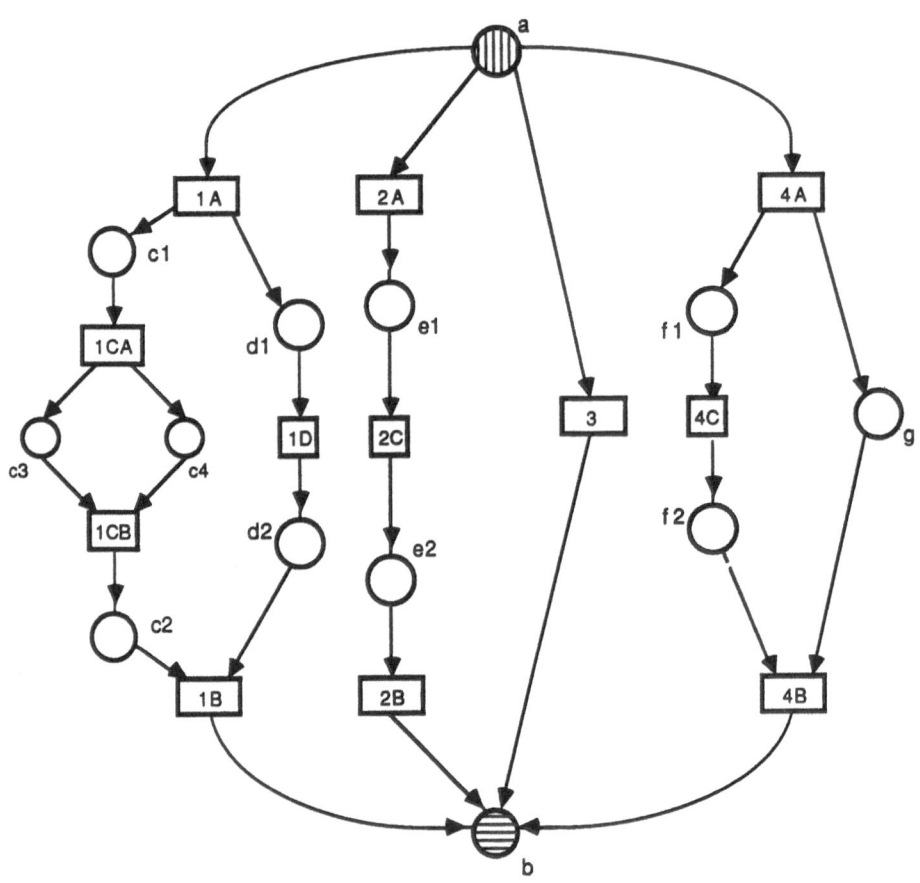

Plan N: Level 4

Fig. 7.5

At level 1, we found that only transition **1** is macro. Our procedure yields:

$$h := 1,$$

$$\mu'_{D1}(1A) = 0.9, \quad \mu'_{D1}(1B) = 0.9, \quad \mu'_{D1}(1D) = 0.5, \quad \mu'_{D1}(1C) = 0.35,$$

$$\mu'_{D1}(1) = \#_D(\{1A, c, 1B\}) = \min\ \{\ \S_D(\{1A, c, 1B\}),\ \S_D(\{1A, d, 1B\})\ \}\ =$$

$$= \min\ \{\ (0.9 + 0.35 + 0.9)/3,\ (0.9 + 0.5 + 0.9)/3\ \} = \min\ \{0.72, 0.77\} = 0.72\ ;$$

and finally:

$$\mu'_D(c_1) = 0.72.$$

Table 7.1

e	$\mu''_D(e)$	$\mu''_K(e)$	e	$\mu''_D(e)$	$\mu''_K(e)$
4000	0.72	0.74	1012	0.58	0.52
3100	0.69	0.67	1003	0.48	0.43
3010	0.74	0.73	0400	0.60	0.46
3001	0.64	0.64	0310	0.65	0.52
2200	0.66	0.60	0301	0.55	0.43
2110	0.71	0.66	0220	0.70	0.58
2101	0.61	0.57	0211	0.60	0.49
2020	0.76	0.72	0202	0.50	0.39
2011	0.66	0.63	0130	0.75	0.64
2002	0.56	0.53	0121	0.65	0.55
1300	0.63	0.53	0112	0.65	0.45
1201	0.58	0.50	0040	0.80	0.7
1120	0.73	0.65	0031	0.70	0.61
1111	0.63	0.56	0022	0.60	0.51
1102	0.53	0.46	0013	0.27	0.42
1030	0.78	0.71	0004	0.40	0.33
1021	0.68	0.62			

The computation of $\mu'_K(c_1)$ proceeds like that of $\mu'_D(c_1)$ but for

$$\mu'_{K3}(1C) = \#_K(\{1CA, 1CB\}) = (\ \S_K(\{1CA, 1CB\}) + \S_K(\{1CA, 1CB\})\ )/2 = 0.35$$

and

$$\mu'_{D1}(1) = \#_D(\{1A, c, 1B\}) = (\ \S_D(\{1A, c, 1B\}) + \S_D(\{1A, d, 1B\})\ )/2 =$$

$$= [\,(0.9 + 0.35 + 0.9)/3 + (0.9 + 0.5 + 0.9)/3\,]\,/\,2 = (0.72 + 0.77)/2 = 0.74$$

which yields

$$\mu'_K(c_1) = 0.74.$$

For the alternative c_4 the same calculations give

$$\mu'_D(c_4) = 0.4 \quad \text{and} \quad \mu'_K(c_4) = 0.33 \;.$$

Now assume that the actual marking of place a is $M(a) = 4$. Let E be the set of executions. The elements of E are multisets

$$e := m_1 c_1 \oplus m_2 c_2 \oplus m_3 c_3 \oplus m_4 c_4$$

with

$$m_1 + m_2 + m_3 + m_4 = M(a) = 4.$$

Two fuzzy attributes over E may then be deduced:

$$A''_D = \{e, \mu''_D(e)\} \quad \text{with membership function}$$

$$\mu''_D(e) := [\, m_1\, \mu'_D(c_1) + m_2\, \mu'_D(c_2) + m_3\, \mu'_D(c_3) + m_4\, \mu'_D(c_4)\,]/4\,,$$

and

$$A''_K = \{e, \mu''_K(e)\} \quad \text{with membership function}$$

$$\mu''_K(e) := [\, m_1\, \mu'_K(c_1) + m_2\, \mu'_K(c_2) + m_3\, \mu'_K(c_3) + m_4\, \mu'_K(c_4)\,]/4\,.$$

The membership functions of these fuzzy attributes are given in Table 7.1, where multisets $m_1 c_1 \oplus m_2 c_2 \oplus m_3 c_3 \oplus m_4 c_4$ were denoted as $m_1\, m_2\, m_3\, m_4$.

In the next sections we shall show how fuzzy attributes over executions can be used to determine the (non-fuzzy) set of optimal executions. This set can be used for decide on an execution. The method is due to Yager [Ya, Zi]. We shall first consider the case that all considered attributes are given the same relevance by the supervisor.

7.4 Preference Degree of Executions. Optimal Executions

Consider any choice schema c, together with its actual marking M.

Let $E = \{e_1, e_2, \dots, e_n\}$ be the actual set of executions of c, and $A = \{A_i\}$ a finite set of fuzzy attributes over E, with $A_i = \{e, \mu_i(e)\}$. We assume that *all* the above attributes express *desirable features* of executions

The real number $\mu_i(e)$ represents the degree to which execution e exhibits attribute A_i.

The intersection Δ of the fuzzy sets A_i is, by definition, a fuzzy set over E too:

$$\Delta := \bigcap_i A_i = \{e, \mu_\Delta(e)\}.$$

The membership degree of execution e in fuzzy set Δ is the smallest membership degree of e in any fuzzy set A_i:

$$\mu_\Delta(e) := \min_i \mu_i(e).$$

$\mu_\Delta(e)$ represents the degree to which execution e possesses the *set* of attributes A. Since the attributes A_i express desirable features of executions, Δ may be interpreted as the fuzzy attribute "preferable". $\mu_\Delta(e)$ represents then the *preference degree* of execution e.

Table 7.2

e	$\mu_\Delta(e)$	e	$\mu_\Delta(e)$
<u>4000</u>	<u>0.72</u>	1012	0.52
3100	0.67	1003	0.43
3010	0.73	0400	0.46
3001	0.64	0310	0.52
2200	0.60	0301	0.43
2110	0.66	0220	0.58
2101	0.57	0211	0.49
<u>2020</u>	<u>0.72</u>	0202	0.39
2011	0.63	0130	0.64
2002	0.53	0121	0.55
1300	0.53	0112	0.45
1210	0.59	0103	0.36
1201	0.50	0040	0.70
1120	0.65	0031	0.61
1111	0.56	0022	0.51
1102	0.46	0013	0.27
1030	0.71	0004	0.33
1021	0.62		

Optimal executions are executions with maximum preference degree. That is: optimal decisions are executions $e' \in E$ such that

$$\mu_\Delta(e') := \max_E \mu_\Delta(e).$$

The fuzzy attribute "preferable" relating to the executions of choice schema 7.5 is given in Table 7.2.

There are two optimal executions: execution 4000, consisting of four executions of alternative c_1, and execution 2020, consisting of two executions of alternative c_1 and two executions of alternative c_3.

Weighted Attributes

When fuzzy attributes of operations are used for making decisions, one will often want to give them different importance. There may be many good reasons for considering the attribute "cheap" more/less important than the attribute "quick" in the framework of a plan execution.

More importance can be given to a fuzzy attribute only by reducing the membership degrees of its elements. Assume the fuzzy attributes "quick" and "cheap" to be associated with a set of operations. I interpret the membership degrees as figures of merit: the higher the membership degree — the cheaper the operation — the better. The same goes for "quick". If "cheap" and "quick" are equally important to me then I can treat the membership degrees as comparable. I might take their average as a figure of merit of a given choice. But suppose that "quick" becomes more important to me than "cheap". Then I will want to be "stricter" about the membership in the fuzzy set of "quick" operations. I will want to substitute the membership degrees of the fuzzy attribute "quick" with those of a fuzzy attribute "very quick".

Since an operation with degree 0.5 for "quick" may have degree 0.1 for "very quick", after this substitution the contribution made to the merit of an operation by its quickness will have fallen, as it should.

The membership degrees in fuzzy attributes are reals between zero and one. Therefore, a natural way to reduce them systematically is to raise them to a positive exponent. We call a fuzzy attribute whose membership function has been raised to a real positive exponent, a *weighted attribute*. This exponent will be called a *weight* of the fuzzy attribute. The greater the weight, the greater the importance associated with the attribute.

For instance, if I express the extreme importance of the attribute "quick" to me by associating the weight 1000 with the fuzzy attribute "quick", then an operation x "quick" with degree 0.5 becomes "quick" with degree 0.5^{1000} to me.

Let $E = \{e_1, e_2, \dots, e_n\}$ be the set of executions of choice schema c, and let $A = \{A_1, A_2, \dots, A_m\} = \{\{e, \mu_i(e)\}\}$ be a set of attributes of over E.

Table 7.3

e	$(\mu_D(e))^{1.5}$	$(\mu_K(e))^{0.5}$	e	$(\mu_D(e))^{1.5}$	$(\mu_K(e))^{0.5}$
4000	0.61	0.86	1012	0.44	072
3100	0.57	0.82	1003	0.33	0.65
3010	0.64	0.85	0400	0.46	0.68
3001	0.51	0.80	0310	0.52	0.72
2200	0.54	0.77	0301	0.41	0.65
2110	0.60	0.81	0220	0.58	0.76
2101	0.48	0.75	0211	0.46	0.70
2020	0.66	0.85	0202	0.35	0.62
2011	0.54	0.79	0130	0.65	0.80
2002	0.42	0.73	0121	0.52	0.74
1300	0.50	0.73	0112	0.52	0.67
1210	0.56	0.77	0103	0.30	0.60
1201	0.44	0.71	0040	0.71	0.83
1120	0.62	0.80	0031	0.58	0.78
1111	0.50	0.75	0022	0.46	0.71
1102	0.38	0.68	0013	0.14	0.65
1030	0.69	0.84	0004	0.25	0.57
1021	0.56	0.79			

A *weighting* of A is vector $w = [w_1, w_2, \dots, w_m]$ of positive reals such that
$$\Sigma_i w_i = m .$$
w_i is called the *weight* of attribute A_i, and the fuzzy sets

$$A_i^* = \{ e, (\mu_i(e))^{w_i} \}$$

are called *weighted attributes* of E.

Now, consider again choice schema 7.5 with it fuzzy attributes **D** and **K**. Assume **w** = [1.5, 0.5] as a weighting of **D** and **K**. This weighting expresses the fact that attribute **K** is given half its usual importance, and that attribute **D** is given 3/2 times its usual importance. We get the weighted attributes

$$D^* = \{ e, (\mu_D(e))^{1.5} \} \quad \text{and} \quad K^* = \{ e, (\mu_K(e))^{0.5} \}$$

of Table 7.3.

Table 7.4

e	$\mu_\Delta(e)$	e	$\mu_\Delta(e)$
4000	0.61	1012	0.44
3100	0.57	1003	0.33
3010	0.64	0400	0.46
3001	0.51	0310	0.52
2200	0.54	0301	0.41
2110	0.60	0220	0.58
2101	0.48	0211	0.46
2020	0.66	0202	0.35
2011	0.54	0130	0.65
2002	0.42	0121	0.52
1300	0.50	0112	0.52
1210	0.56	0103	0.30
1201	0.44	<u>0040</u>	<u>0.71</u>
1120	0.62	0031	0.58
1111	0.50	0022	0.46
1102	0.38	0013	0.14
1030	0.69	0004	0.25
1021	0.56		

Observe that a different weighting of attributes **D** and **K** with the same proportion of weights — for instance, weighting **w'** = [3, 1] — would express a different "absolute importance", but the same "relative importance" of attributes **D** and **K**. Weighting **w'** says that attribute **K** is given "the same the importance as usual", and that — again — attribute **D** is given three times the importance of **K**.

The preference degree of executions is given in Table 7.4, and the optimal execution turns out to be execution 0040, consisting of four executions of alternative c_3.

Observe that a different weighting of attributes **D** and **K** that conserves the proportion of weights — as **w'** = [3, 1] does — would of course change the preference degrees listed in Table 7.4, but it would not change their ordering. Indeed, for all $k > 0$

$$a^m \geq b^n \quad \rightarrow \quad a^{km} \geq b^{kn}.$$

Attribute weights may be initially fixed by the supervisor — according to his general beliefs and interests, or assigned during the actual plan execution. Sometimes, evaluating the importance of attributes to one another by pair-wise comparison is easier than assigning the weights directly. *Saaty's method* [Sa] allows one to deduce the weighting of an attribute set from the reciprocal weight of attributes.

Let $A = \{ A_1, A_2, \ldots, A_n \}$ be a set of attributes, and denote by $w = [w_1, w_2, \ldots, w_n]$ a weighting of A. Suppose that the relative importance of attribute A_i to attribute A_j is expressed by a positive real number r_{ij}. We call the $n \times n$ matrix **R** whose generic element is r_{ij} the *matrix of relative weights* of A.

For i, j = 1, 2, ... , n we have

$$r_{ij} = r_{ji}^{-1} \quad \text{and} \quad r_{ii} = 1.$$

We say that **R** is *consistent* if

$$r_{ik}\, r_{kj} = r_{ij}.$$

A non-consistent matrix of relative weights represents non-consistent judgments about the relative importance of attributes.

If matrix **R** is consistent, we can assume that $r_{ij} = w_i / w_j$. That is, that

$$
\mathbf{R} =
\begin{bmatrix}
w_1/w_1 & w_1/w_2 & \cdots & w_1/w_n \\
w_2/w_1 & w_2/w_2 & \cdots & w_2/w_n \\
& & & \\
w_n/w_1 & w_n/w_2 & \cdots & w_n/w_n
\end{bmatrix}
$$

In this case, the right multiplication of R by w^T will give the vector nw^T. Therefore, if R is consistent, the weighting w is a solution of the linear system

$$R\ w^T = nw^T.$$

and the required weighting w only exists if n is an eigenvalue of matrix R.

This can be guaranteed. Since all rows of R are non-zero, and proportional to one another, R has rank one. Therefore, its n eigenvalues $\lambda_1, \lambda_2, \ldots, \lambda_n$ are all equal to zero except one, λ^*.

We know that the sum of the eigenvalues of R equals the sum of the elements of the main diagonal. Therefore $\Sigma_i \lambda_i = \lambda^* = n$, and n is the only positive eigenvalue of R.

Any column of R is an eigenvector corresponding to the eigenvalue n. We will choose the eigenvector w normalized by $\Sigma_i w_i = 1$ as the weighting of attribute set A.

If matrix R is not consistent — as will usually be the case — the above procedure does not work. Then λ^* will be different from n. But it is known that in any matrix small perturbations in the entries result in small perturbations in the eigenvalues. By the Perron-Frobenius theorem, every matrix with positives entries has a unique real positive eigenvalue λ^*, and the normalized eigenvector w^* corresponding to λ^* has non-negative entries.

If R is "consistent enough" — that is, if it is "close enough" to a consistent matrix of relative weights — we can take w^* as the weighting of our attribute set A.

Returning to attributes D and K of the previous example, we could for instance assign matrix

	D	K
D	1	4
K	1/4	1

as the matrix of relative weights. This matrix has exactly one positive eigenvalue: $\lambda^* = 2$. This yields one eigenvector w^* such that $\Sigma_i w_i = 1$, the eigenvector

$$w^* = [0.8, 0.2].$$

w^* assigns a weighting of attribute set $\{D, K\}$ consistent with the pairwise comparison of attributes D and K expressed by the above matrix of relative weights. Assigning

	D	K
D	1	3
K	1/3	1

as the matrix of relative weights, we would obtain the weighting
$$\mathbf{w}^* = [0.75, 0.25].$$

As a last example, consider a choice schema $\mathbf{c}$ with five executions: e_1, e_2, e_3, e_4 and e_5. Also, we assume that four fuzzy attributes are given over the set of executions: $A_i = \{ e, \mu_i(e) \}$ with $i = 1, 2, 3, 4$.

Table 7.5

e	$\mu_1(e)$	$\mu_2(e)$	$\mu_3(e)$	$\mu_4(e)$
e_1	0.25	0.50	0.70	0.30
e_2	0.35	0.65	0.80	0.50
e_3	0.25	0.70	0.95	0.35
e_4	0.90	0.40	0.15	0.80
e_5	0.10	0.70	0.70	0.90

Let the membership functions of the A_i be those of Table 7.5, and let a pair-wise comparison of the four attributes be provided by the following matrix $\mathbf{R}$:

	A_1	A_2	A_3	A_4
A_1	1	3	2	5
A_2	1/3	1	2	1
A_3	1/2	1/2	1	5
A_4	1/5	1	1/5	1

$\mathbf{R}$ is not consistent since, for instance, $r_{2,1}r_{1,4} \neq r_{2,4}$. The eigenvalue of $\mathbf{R}$ which is closer to 4 is $\lambda^* = 4.492$, and yields the normalized eigenvector
$$\mathbf{w}^* = [0.45, 0.21, 0.24, 0.10].$$

We will take $\mathbf{w}^*$ as a weighting of attribute set A resulting from the comparison of attributes given by matrix $\mathbf{R}$.

Table 7.6

e	$(\mu_1(e))^{0.45}$	$(\mu_2(e))^{0.21}$	$(\mu_3(e))^{0.24}$	$(\mu_4(e))^{0.10}$
e_1	0.53	0.86	0.92	0.87
e_2	0.62	0.91	0.95	0.93
e_3	0.53	0.93	0.99	0.90
e_4	0.95	0.82	0.63	0.98
e_5	0.35	0.93	0.92	0.99

Table 7.7

e	$\mu_\Delta(e)$
e_1	0.53
e_2	0.62
e_3	0.53
e_4	0.63
e_5	0.35

The membership functions of the weighted attributes are those in Table 7.6. The preference degree of executions is given in Table 7.7. Execution e_4 is optimal, since for it $\mu_\Delta(e_4) = 0.63 = \max_E \mu_\Delta(e)$.

If the pair-wise comparison of our four attributes was provided by the following matrix $\mathbf{R}'$

	A_1	A_2	A_3	A_4
A_1	1	1	2	2
A_2	1	1	8	9
A_3	1/2	1/8	1	7
A_4	1/2	1/9	1/7	1

then the required weighting of our attribute set would be given by eigenvector $\mathbf{w}^* =$ [0.25, 0.53, 0.15, 0.07]. The membership functions of the weighted attributes would be those of Table 7.8, and e_2 would turn out to be the optimal execution.

Table 7.8

e	$(\mu_1(e))^{0.25}$	$(\mu_2(e))^{0.53}$	$(\mu_3(e))^{0.15}$	$(\mu_4(e))^{0.07}$
e_1	0.71	0.69	0.95	0.92
e_2	0.77	0.79	0.97	0.95
e_3	0.71	0.83	0.99	0.93
e_4	0.97	0.61	0.75	0.98
e_5	0.56	0.83	0.95	0.99

7.5 Clustering Executions into Preference Classes: Fuzzy Outranking

We have seen how optimal executions of choice schemata can be deduced from a set of fuzzy operation attributes. The preference degree of executions was used to single out optimal executions: an execution is *strictly preferred* to another if and only if its preference degree is greater. Two executions with the same preference degree are considered to be *indifferent* to the supervisor. The desirability of executions is assumed to be always comparable.

Sometimes, considering *incomparable* executions may also be useful. Moreover, a supervisor might need preference categories other than strict preference and indifference alone. The choice problems associated with project plans often require the definition of more specific preference classes.

Fuzzy outranking was introduced by B. Roy [Ro] to meet these needs. This technique allows partitioning executions into preference classes on the basis of their reciprocal *outranking degree* — where the *outranking degree of execution* A *over execution* B represents the plausibility of the statement "The supervisor will find A at least as good as B". The preference classes are ranked according to some preference scale. These preference classes and their relations to one another provide a model of the preference structure of the supervisor.

As we will see, fuzzy outranking allow us to determine the class of best executions, but also the classes of second-best executions, undetermined executions, and others to be defined according to the specific choice problems posed by a particular application.

Fuzzy Outranking Relation

Consider the execution of a plan with alternatives. In order to steer the course of action, in order to make decisions, an execution supervisor has to establish a scale of merit between executions. He will consider some executions indifferent to others, and some executions better, or decisively better, than others.

Fuzzy relations can be used for formalizing the *supervisor's view* on whether and how much an execution outranks another execution. If merits of executions are expressed as fuzzy attributes, their concordance/discordance with the supervisor's beliefs can be evaluated on the basis of their membership functions.

The *outranking degree of execution* e_h *over execution* e_k is computed by aggregating concordance and discordance factors. It defines the *fuzzy outranking relation*.

Consider again a choice schema c, together with the set $E = \{e_1, e_2, \dots, e_n\}$ of its executions. Let $A_i = \{e, \mu_i(e)\}$ be m fuzzy attributes over E, and $w_i \geq 0$ their relative weights, with $i = 1, 2, \dots, m$, and $\Sigma_i w_i = 1$.

We assume that the supervisor specified, for each attribute A_i, three real values
the *indifference threshold* I_i,
the *preference threshold* P_i,
the *veto threshold* V_i,
such that $0 < I_i < P_i < V_i < 1$, with the following interpretation:

e_h is as good as e_k iff $\mu_i(e_h) - \mu_i(e_k) \geq I_i$,

e_h is preferred to e_k iff $\mu_i(e_h) - \mu_i(e_k) \geq P_i$,

e_h is considerably better than e_k iff $\mu_i(e_h) - \mu_i(e_k) \geq V_i$.

On the basis of these thresholds, the *concordance matrix of attribute* A_i

$$C^i = [c^i_{hk}]$$

is defined, the generic element of which expresses the plausibility of the statement "Execution e_h is at least as good as execution e_k with respect to attribute A_i". We set:

$$c^i_{hk} := \begin{cases} 1 & \text{if} \quad \mu_i(e_k) \leq \mu_i(e_h) + I_i \\ [\mu_i(e_h) - \mu_i(e_k) + P_i)] / (I_i - P_i) & \text{if} \quad \mu_i(e_h) + I_i \leq \mu_i(e_k) \leq \mu_i(e_h) + P_i \\ 0 & \text{if} \quad \mu_i(e_k) \geq \mu_i(e_h) + P_i \end{cases}$$

Similarly, we define the *discordance matrix of attribute* A_i

$$D^i = [d^i_{hk}]$$

the generic element of which expresses the plausibility of the statement "Execution e_h is not at least as good as execution e_k with respect to attribute A_i". We set:

$$d^i_{hk} := \begin{cases} 0 & \text{if} \quad \mu_i(e_k) \leq \mu_i(e_h) + P_i \\ [\mu_i(e_k) - \mu_i(e_h) + P_i)] / (V_i - P_i) & \text{if} \quad \mu_i(e_h) + P_i \leq \mu_i(e_k) \leq \mu_i(e_h) + V_i \\ 1 & \text{if} \quad \mu_i(e_k) \geq \mu_i(e_h) + V_i \end{cases}$$

The concordance and discordance matrices of attributes must be aggregated in order to get the outranking relation. Following Roy's approach, we first determine the *total concordance matrix* C, the weighted sum of the concordance matrices of all attributes:

$$C = [c_{hk}] := \Sigma_i w_i C^i.$$

Concordance factors then have to be aggregated with discordance factors. This is achieved by means of the *discordance multiplier* d_{hk}. This multiplier is defined by

$$d_{hk} := (\Sigma_i d^i_{hk} \bullet c_{hk}) / m$$

where

$$d^i_{hk} \bullet c_{hk} := \begin{cases} 1 & \text{if} \quad d^i_{hk} \leq c_{hk} \\ (1 - d^i_{hk})/(1 - c_{hk}) & \text{if} \quad d^i_{hk} > c_{hk} \end{cases}$$

(observe that $d^i_{hk} > c_{hk}$ implies $1 - c_{hk} \neq 0$).

The *degree of outranking of execution* e_h *over execution* e_k is, hence, defined as

$$r_{hk} := c_{hk} d_{hk},$$

and directly yields the *fuzzy outranking relation over* E

$$R := \{ ((e_h, e_k), \mu_R(e_h, e_k)) \mid (e_h, e_k) \in E \times E \ \wedge \ \mu_R(e_h, e_k) = r_{hk} \}.$$

We will illustrate this technique by again considering the choice schema **c** and the four fuzzy attributes $A_i = \{e, \mu_i(e)\}$ of Table 7.5. We take values of Table 7.9 as the indifference, preference and veto thresholds of the various attributes.

Table 7.9

	A_1	A_2	A_3	A_4
I_i	0.1	0.1	0.05	0.1
P_i	0.2	0.2	0.1	0.01
V_i	0.22	0.4	0.3	0.9

Tables 7.10 to 7.13 show the concordance matrices of attributes A_i. The reader is encouraged to observe how the assigned threshold values influence the evaluation of the concordance between attributes.

Table 7.10

C^1	e_1	e_2	e_3	e_4	e_5
e_1	1	1	1	0	1
e_2	1	1	1	0	1
e_3	1	1	1	0	1
e_4	1	1	1	1	1
e_5	0.5	0	0.5	0	1

Table 7.11

C^2	e_1	e_2	e_3	e_4	e_5
e_1	1	0.5	0	1	0
e_2	1	1	1	1	1
e_3	1	1	1	1	1
e_4	1	0	0	1	0
e_5	1	1	1	1	1

Table 7.12

C^3	e_1	e_2	e_3	e_4	e_5
e_1	1	0	0	1	1
e_2	1	1	0	1	1
e_3	1	1	1	1	1
e_4	0	0	0	1	0
e_5	1	0	0	1	1

Table 7.13

C^4	e_1	e_2	e_3	e_4	e_5
e_1	1	0.66	1	0	0
e_2	1	1	1	0.33	0
e_3	1	0.83	1	0	0
e_4	1	1	1	1	1
e_5	1	1	1	1	1

If $w = [1/8, 1/2, 1/8, 1/4]$ is the weighting of attributes A_i, the total concordance matrix C is that of Table 7.14. The entries of the row corresponding to execution e_3 indicate the plausibility degree of the statement "Execution e_3 is at least as good as execution e_k, given our attributes, thresholds and weighting".

Table 7.14

C	e_1	e_2	e_3	e_4	e_5
e_1	1	0.54	0.37	0.75	0.25
e_2	1	1	0.87	0.71	0.75
e_3	1	0.96	1	0.62	0.75
e_4	0.87	0.37	0.37	1	0.37
e_5	0.94	0.75	0.81	0.87	1

Tables 7.15 to 7.18 show the concordance matrices of attributes A_i. The reader is encouraged to observe the influence of the assigned threshold values on the evaluation of the discordance of attributes.

Table 7.15

D^1	e_1	e_2	e_3	e_4	e_5
e_1	0	0	0	1	0
e_2	0	0	0	1	0
e_3	0	0	0	1	0
e_4	0	0	0	0	0
e_5	0	1	0	1	0

Table 7.16

D^2	e_1	e_2	e_3	e_4	e_5
e_1	0	0	0	0	0
e_2	0	0	0	0	0
e_3	0	0	0	0	0
e_4	1	0	0	1	0
e_5	0	0.25	0.5	0	0.5

Table 7.17

D^3	e_1	e_2	e_3	e_4	e_5
e_1	0	0	0.25	0	0
e_2	0	0	0.25	0	0
e_3	0	0	0	0	0
e_4	1	1	1	0	1
e_5	0	0	0.25	0	0

Table 7.18

D^4	e_1	e_2	e_3	e_4	e_5
e_1	0	0.21	0.04	0.55	0.66
e_2	0	0	0	0.32	0.44
e_3	0.16	0	0	0.49	0.61
e_4	0	0	0	0	0.10
e_5	0	0	0.25	0	0

The computation of the discordance multipliers requires the determination of matrices $[d^i_{hk} \bullet c_{hk}]$. They are given in Tables 7.19 to 7.22.

Table 7.19

$[d^1_{hk} \bullet c_{hk}]$	e_1	e_2	e_3	e_4	e_5
e_1	1	1	1	0	1
e_2	1	1	1	0	1
e_3	1	1	1	0	1
e_4	1	1	1	1	1
e_5	1	0	1	0	1

Table 7.20

$[d^2_{hk} \bullet c_{hk}]$	e_1	e_2	e_3	e_4	e_5
e_1	1	1	1	1	1
e_2	1	1	1	1	1
e_3	1	1	1	1	1
e_4	0	1	1	0	1
e_5	1	1	1	1	1

Table 7.21

$[d^3_{hk} \bullet c_{hk}]$	e_1	e_2	e_3	e_4	e_5
e_1	1	1	1	1	1
e_2	1	1	1	1	1
e_3	1	1	1	1	1
e_4	0	0	0	1	0
e_5	1	1	1	1	1

Table 7.22

$[d^4_{hk} \cdot c_{hk}]$	e_1	e_2	e_3	e_4	e_5
e_1	1	1	1	1	0.45
e_2	1	1	1	1	1
e_3	1	1	1	1	1
e_4	1	1	1	1	1
e_5	1	1	1	1	1

The discordance multipliers d_{hk} are listed in Table 7.23. They were computed, by definition, as $d_{hk} := (d^1_{hk} \cdot c_{hk} + d^2_{hk} \cdot c_{hk} + d^3_{hk} \cdot c_{hk} + d^4_{hk} \cdot c_{hkk})/4.$

Finally, we compute matrix $[r_{hk}] = [c_{hk} d_{hk}]$ representing the fuzzy outranking relation **R**. This matrix is shown in Table 7.24.

Table 7.23

$[d_{hk}]$	e_1	e_2	e_3	e_4	e_5
e_1	1	1	1	0.75	0.86
e_2	1	1	1	0.75	1
e_3	1	1	1	0.75	1
e_4	0.5	0.75	0.75	0.75	0.75
e_5	1	0.75	1	0.75	1

Table 7.24

$[r_{hk}]$	e_1	e_2	e_3	e_4	e_5
e_1	1	0.79	0.37	0.46	0.21
e_2	1	1	0.87	0.53	0.75
e_3	1	0.96	1	0.46	0.75
e_4	0.43	0.28	0.28	0.75	0.28
e_5	0.93	0.56	0.81	0.65	1

Classification of Executions

We will now show how the fuzzy outranking relation can be used to classify executions of choice schemata into preference classes, and to rank these classes into a desirability scale. These classes represent the preference categories associated by the decision maker/supervisor with the alternative executions of the considered plan.

For some plans, it will be sufficient to consider the class of best executions. For other plans, considering other preference classes may be necessary, such as the class of second-best executions, or that of undetermined ones.

As we will show, the fuzzy outranking relation can be used [Ro, RV] both for partitioning executions into *two* preference classes *(dichotomy)*, or into *several* preference classes *(polychotomy)*. In the first case, the class of best executions will be singled out; in the second, executions will be partitioned into the several preference classes established by the supervisor, and by construction ranked into a preference ordering.

Again, c will be a choice schema, $E = \{e_1, e_2, \dots, e_n\}$ its set of executions, $\{A_i\}$ a finite set of fuzzy attributes over E, and $w_i \geq 0$ their weights.

We assume that for each attribute A_i the indifference, preference threshold, and veto threshold are specified, and that the overall fuzzy outranking relation $R = \{ (e_h, e_k), \mu_R (e_h, e_k)\}$ has been determined.

Singling out Best Executions: Dichotomy

From fuzzy outranking relation R an ordinary relation over set of executions E is derived, the outranking relation R^α.

Given a real number $\alpha \in [0, 1]$, the *outranking relation* R^α is defined as the α-*cut* of R. That is:

$$R^\alpha := \{ (e_h, e_k) \mid (e_h, e_k) \in E \times E \ \wedge \ r_{hk} \geq \alpha \}.$$

If there exist executions e' such that

$$\forall e_k \in E: \ (e', e_k) \in R^1,$$

these are the *best executions*.

Otherwise — as will more often be the case — the supervisor will have to look for the greatest α such that

$$\exists e' \in E \mid \ \forall e_k \in E : \ (e', e_k) \in R^\alpha.$$

If this α is judged high enough to claim that e' dominates the remaining e_k, e' will be considered a best execution.

The fuzzy outranking relation of Table 7.24, for instance, yields 0.56 as the greatest α with the above property, and e_5 as the corresponding execution. If the supervisor considers $\alpha = 0.56$ a plausible dominance degree for best executions, then e_5 will be a best execution — the only one.

By definition there may be several best executions. If so, the supervisor chooses the one to execute without numeric guidance.

Partitioning Executions into Preference Classes: Polychotomy

Executions are partitioned into preference classes using their outranking degree over other executions, and a finite succession of preference thresholds. To this end, the planner will specify a finite succession of reals

$$0 = \lambda_1 \leq \lambda_2 < \lambda_3 < \ldots\ldots < \lambda_n \leq \lambda_{n+1} = 1,$$

to be interpreted as preference thresholds.

Each interval $[\lambda_i, \lambda_{i+1}]$ is interpreted as a preference interval, and an execution e_h is classified as belonging to the *i-th preference class* if, and only if,

$$\min_k r_{hk} \geq \lambda_i \quad \wedge \quad \max_k r_{hk} \leq \lambda_{i+1}.$$

All executions not belonging to the preference classes are classified as *undetermined*. We define the set of undetermined executions to be the *0-th preference class*.

Preference classes are normally given an interpretation. For instance, executions can be classified as either "first-choice", or "second-choice", or "third-choice", or "undetermined" by assigning two preference thresholds, λ_1 and λ_2, such that $0 \leq \lambda_1 \leq \lambda_2 \leq 1$. Execution e_h will then be classified as:

- "first-choice", if its outranking degree over any other execution is greater than λ_2. That is, if:

$$\min_k r_{hk} \geq \lambda_2.$$

- "Second-choice", if its outranking degree over any other execution belongs to the interval $[\lambda_1, \lambda_2]$. That is, if:

$$\lambda_1 \leq \min_k r_{hk} \leq \max_k r_{hk} \leq \lambda_2.$$

- "Third-choice", if its outranking degree over any other execution is less than λ_1. That is, if:

$$\max_k r_{hk} \leq \lambda_1.$$

- "Undetermined", otherwise.

Table 7.25

$[r_{hk}]$	e_1	e_2	e_3	e_4	e_5	e_6
e_1	0.50	0.79	0.37	0.46	0.48	0.70
e_2	0.33	0.32	0.27	0.13	0.17	0.25
e_3	0.43	0.28	0.28	0.75	0.20	0.55
e_4	0.75	0.78	0.98	0.75	0.80	1
e_5	1	0.75	0.81	0.76	0.88	0.95
e_6	0.50	0.75	0.41	0.60	0.48	0.85

For $\lambda_1 = 0.35$ and $\lambda_2 = 0.70$, the fuzzy outranking relation given in Table 7.25 yields the following classification of executions: e_4 and e_5 "first-choice", e_1 "second-choice", e_2 "third-choice", e_3 and e_6 "undetermined".

Setting $\lambda_1 = 0.30$ while keeping $\lambda_2 = 0.70$ changes the classification of both e_1 and e_2 to "undetermined".

The values assigned to the λ_i determine the outcome of the polychotomy. Their choice adds information to the plan, information the origin and nature of which is not taken into account within the planning framework we have presented.

References

[BP]   Burgess R.R., Pritsker A.A.B.: The GERTS Simulation Programs: GERTS III, GERTS III Q, GERTS III C, GERTS III. Electronic Research Centre, NASA

[BS]   Best E., Schmid H.: Towards a Constructive Solution of the Liveness Problem in Petri Nets. Technical Report 4/76, Institut für Informatik, Universität Stuttgart, 1976

[CM1] Clayton E.R., Moore L.J.: GERT Network Simulation. Proc. of the 3rd Annual Meeting of the Southeastern Division of the American Institute of Decision Sciences, 1973

[CM2] Clayton E.R., Moore L.J.: GERT Modeling and Simulation: Fundamentals and Applications. Petrocelli Charter, New York 1976

[CS]   Colom J.M., Silva M.: Convex Geometry and Semiflows in P/T Nets. A Comparative Study of Algorithms for Computation of Minimal P-Semiflows. In: Proceedings of the 10th International Conference on Application and Theory of Petri Nets, Bonn 1989

[De]   De Ambrogio W.: Programmazione reticolare, Vol.I. Etas Libri, Milan 1977

[Fi]   Fisz M.: Probability Theory and Mathematical Statistics. John Wiley and Sons, New York, London 1963

[FJ]   Feldbrugge F., Jensen K.: Petri Net Tool Overview 1986. In: W. Brauer et al.

(eds.): Petri Nets: Applications and Relationships to Other Models of Concurrency. Lecture Notes in Computer Science, Vol. 255, Springer-Verlag, Berlin, Heidelberg, New York, 1987

[Fo] Fourier J.B.J.: Solution d'une question particulière du calcul des inegalités. In: Oeuvres II, pp. 317-328; Gauthier-Villars, Paris

[Gr] Grubbs E.F.: Attempts to Validate Certain PERT Statistics. Operations Research, **10:** 6, 1962

[JK] Jaxy M., Krückeberg F.: Mathematical Methods for Calculating Invariants in Petri Nets. In: G. Rozenberg (ed.): Advances in Petri Nets 1987. Lecture Notes in Computer Science, Vol. 266, Springer- Verlag, Springer-Verlag, Berlin, Heidelberg, New York, 1987

[La] Lautenbach K.: Liveness in Petri Nets. Internal Report GMD-ISF 72-O2.1, sankt Augustin, 1972

[Li] Lipton J.R.: The Reachability Problem Requires Exponential Space. Yale Univ., Dept. of Comp. Sci., Research Report No. 62, 1976

[Lie] Lien Y.: Termination Properties in Generalized Petri Nets. SIAM Journal of Computing **5:** 2, 1976, 251-265

[MM] Mayr E.W., Meyer A.R.: The Complexity of the Finite Containment Problem for Petri Nets. Journal of the ACM **28:** 3, 1981, 561-576

[MP] Moder J.J., Phillips C.R.: Project Management with CPM and PERT. Van Nostrand Reinhold, New York, 1970

[NS] Neumann K., Steihard U.: GERT Networks and the Time-Oriented Evaluation of Projects. Springer-Verlag, Berlin, Heidelberg, New York, 1979

[MR] MacCrimmon K.R.: Ryavec C.A., An Analytical Study of the PERT Assumptions. Operations Research **12:** 1, 1964

[Ro] Roy B.: Partial Preference Analysis and Decision-Aid: the Fuzzy Outranking Relation Concept. In: SEMA, Paris, 1976

[RV] Roy B.: Vinke Ph., Multicriteria Analysis: Survey and New Directions. EJOR 8, 207-218

[Sa] Saaty T.L.: Exploring the Interface between Hierarchies, Multiple Objectives and Fuzzy Sets. Fuzzy Sets and Systems 1, 1978, 57-68

[Vi]   Vianelli S.: Prontuari per calcoli statistici. Abbaco, Palermo, Roma, 1959

[Ya]   Yager R.R.: Fuzzy Decision Making Including Inequal Objectives. Fuzzy Sets and Systems 1, 1978, 87-95

[Zi]   Zimmermann H.J.: Fuzzy Sets, Decision Making, and Expert Systems. Kluwer Academic Publishers, Dordrecht, Boston, London, 1987

Further Reading

This section is aimed at helping the reader deepen the understanding of the notions and techniques presented in this book. The proposed readings are comprehensive works on general or special topics presented in the book. For the reader's convenience, we have avoided suggesting conference proceedings or scattered papers. The selection mirrors the author's personal taste and experience as a reader. It does not pretend to be complete, but should certainly help the interested traveller along the way.

Graph Theory

L. W. Beinecke, R. J. Wilson, *Applications of Graph Theory,* Academic Press, New York, 1979

N. Deo, *Graph Theory with Applications to Engineering and Computer Science,* Prentice-Hall, Englewood Cliffs, New Jersey, 1974

F. Harary, *Graph Theory,* Addison-Wesley, Reading, Mass., 1969

F. Harary, R. Z. Norman, D. Cartwright, *Structural Models: An Introduction to the Theory of Directed Graphs,* John Wiley and Sons, New York, 1965

R. J. Wilson, *Introduction to Graph Theory,* Academic Press, New York, 1972, and Longman Group, Harlow, Essex, 1975

Networking Techniques

A. Alan, B. Pritsker, *Modeling and Analysis Using Q-GERT Networks,* John Wiley and
Sons, New York, 1979

R. D. Archibald, R. L. Villoria, *Network-Based Management Systems (PERT/CPM),*
John Wiley and Sons, New York, 1967

D. D. Bedworth, *Industrial Systems, Planning, Analysis, Control,* Ronald Press, New
York, 1973

E. S. Buffa, W. H. Taubert, *Production-Inventory Systems: Planning and Control,*
Irwin, Homewood, Ill., 1972

E. R. Clayton, L. J. Moore, *GERT Modeling and Simulation: Fundamentals and
Applications,* Petrocelli Charter, New York, 1976

S. E. Elmaghraby, *Activity Networks: Project Planning and Control by Network Models,*
John Wiley and Sons, New York, 1977

C. A. Kirkpatrick, R. Levin, *Planning and Control with PERT/CPM,* McGraw-Hill,
New York, 1966

R. F. Gonzales, C. McMillan, *Systems Analysis, A Computer Approach to Decision
Models,* (3rd ed.), Irwin, Homewood, Ill., 1973

R. W. Miller, *Schedule, Cost and Profit Control with PERT,* McGraw-Hill, New York,
1963

J. J. Moder, C. R. Phillips, *Project Management with CPM and PERT,* Van Nostrand
Reinhold, New York, 1970

K. Neumann, U. Steihard, *GERT Networks and the Time-Oriented Evaluation of
Projects,* Springer-Verlag, Berlin, Heidelberg,1979

G. E. Whitehouse, *System Analysis and Design Using Network Techniques,* Prentice-
Hall, New York, 1973

Petri Nets

G.W. Brams (nom collectiv), *Réseaux de Petri: Théorie et Practique,* Tomes 1 et 2,
Editions Masson, Paris, 1982

W. Brauer, G. Rozenberg, W. Reisig (eds.), *Petri Nets: Central Models and their Properties*, Lecture Notes in Computer Science, Vol.254, Springer-Verlag, Berlin, Heidelberg, New York, 1987

W. Brauer, G. Rozenberg, W. Reisig, (eds.), *Petri Nets: Applications and Relationships to Other Models of Concurrency*, Lecture Notes in Computer Science, Vol.255, Springer-Verlag, Berlin, Heidelberg, New York, 1987

W. Reisig, *Petri Nets, An Introduction*, Springer-Verlag, Berlin, Heidelberg, New York, 1985

W. Reisig, *Systementwurf mit Netzen*, Springer-Verlag, Berlin, Heidelberg, New York, 1985

M. Silva, *Las Redes de Petri: en la Automática y la Informática*, Editorial AC, Madrid, 1985

G. Rozenberg, (ed.), *Advances in Petri Nets 1987*, Lecture Notes in Computer Science, Vol.266, Springer-Verlag, Berlin, Heidelberg, New York, 1987

G. Rozenberg, (ed.), *Advances in Petri Nets 1988*, Lecture Notes in Computer Science, Vol.340, Springer-Verlag, Berlin, Heidelberg, New York, 1988

Decision Support Systems

S. S. Mitra, *Decision Support Systems — Tools and Techniques*, John Wiley and Sons, New York, 1986

R. J. Thierauf, *Decision Support Systems for Effective Planning and Control*, Prentice-Hall, Englewood Cliffs, New Jersey, 1982

Multicriteria Decision Making

K. J. Arrow, *Individual Choice under Certainty and Uncertainty*. In: Collected Papers of K. J. Arrow, Blackwell, Oxford, 1984

D. Bell, R. Keeney, H. Raiffa (eds.), *Conflicting Objectives in Decisions*, John Wiley and Sons, New York, 1977

A. H. Cornell, *The Decision-Maker's Handbook*, Prentice-Hall, Englewood Cliffs, 1980

G. Fandel, J. Spronk, *Multiple Criteria Decision Methods and Applications,* Springer-Verlag, Berlin, Heidelberg, New York, 1985

G. Fandel, *Optimale Entscheidungen in Organisationen,* Springer-Verlag, Berlin, Heidelberg, New York, 1979

N. A. J. Hastings, J. M. C. Mello, *Decision Networks,* John Wiley and Sons, New York, 1979

R. Keeney, H. Raiffa, *Decisions with Multiple Objectives: Preferences and Value Trade-offs,* John Wiley and Sons, New York, 1976

D. W. Miller, M. K. Starr, *Executive Decisions and Operations Research,* Prentice-Hall, Englewood Cliffs, 1960

J. de Montgolfier, P. Bertier, *Approche multicritère des problèmes des décision,* Editions Hommes et Techniques, Paris, 1978

Saaty T. L., *The Analytic Hierarchy Process. Planning, Priority Setting and Resource Allocation,* McGraw-Hill, New York, 1980

H. A. Simon, *Models of Bounded Rationality,* The MIT Press, Cambridge, Mass., 1982

H. Thiriez, S. Zionts (eds.), *Multiple Criteria Decision Making,* Lecture Notes in Economics and Mathematical Systems, Vol.30, Springer-Verlag, Berlin, Heidelberg, New York, 1976

M. Zeleny, *Multiple Criteria Decision Making,* McGraw Hill, New York, 1981

Fuzzy Sets

T. T. Ballmer, M. Pinkal, *Approaching Vagueness,* North-Holland, New York, 1983

D. Dubois, H. Prade, *Fuzzy Sets and Systems: Theory and Applications,* Academic Press, New York, 1980

T. A. Folger, G. J. Klir, *Fuzzy Sets, Uncertainty, and Information,* Prentice Hall, Englewood Cliffs, New Jersey, 1988

A. Kaufmann, *Introduction to the Theory of Fuzzy Subsets,* Academic Press, New York, 1975

M. M. Gupta, A. Kaufmann, *Introduction to Fuzzy Arithmetic: Theory and Applications,* Van Nostrand Reinhold, New York, 1985

M. M. Gupta, R. K. Ragade, R. R. Yager (eds.), *Advances in Fuzzy Set Theory and Applications,* North-Holland, New York, 1979

H. J. Zimmermann, *Fuzzy Set Theory — and its Applications,* Kluwer-Nijhoff, Boston, 1985

Fuzzy Decision Making

K. Fu, M. Shimura, K. Tanaka, L. A. Zadeh (eds.), *Fuzzy Sets and Their Application to Cognitive and Decision Processes,* Academic Press, New York, 1975

B. R. Gaines, H. J. Zimmermann, L. A. Zadeh (eds.), *Fuzzy Sets and Decision Analysis,* North-Holland, New York, 1984

W. J. M. Kickert, *Fuzzy Theories on Decision Making,* Martinus-Nijhoff, Boston, 1978

H. J. Zimmermann, *Fuzzy Sets, Decision Making, and Expert Systems,* Kluwer Academic Publishers, Boston, Dodrecht, Lancaster, 1987

Miscellanea

B. De Finetti, *Theory of Probability,* John Wiley and Sons, New York, 1974

M. Fisz, *Probability and Mathematical Statistics,* John Wiley and Sons, New York, 1963

W. Feller, *An Introduction to Probability Theory and Its Applications,* Vols. 1 and 2, John Wiley and Sons, New York, 1950, 1966

R. A. Howard, *Dynamic Probabilistic Systems,* Vols. 1 and 2, John Wiley and Sons, New York, 1971

Index